Additive Manufacturing Technology

This series systematically summarizes the technology developments in the additive manufacturing field in China in recent years, introducing the technical development status in terms of additive manufacturing processes, materials, technologies, and applications.

This series is one of the national key publishing projects in China, and has been listed in the national key book projects of China's "13th Five-Year Plan", supported by the National Publishing Fund.

Weijun Zhu · Dichen Li

Models for Wind Tunnel Tests Based on Additive Manufacturing Technology

国防工业出版社
National Defense Industry Press

Springer

Weijun Zhu
School of Mechanical Engineering
and Automation
Beihang University
Beijing, China

Dichen Li
School of Mechnical Engineering
Xi'an Jiaotong University
Xi'an, Shaanxi, China

ISSN 2731-6114 ISSN 2731-6122 (electronic)
Additive Manufacturing Technology
ISBN 978-981-99-5876-4 ISBN 978-981-99-5877-1 (eBook)
https://doi.org/10.1007/978-981-99-5877-1

Jointly published with National Defense Industry Press
The print edition is not for sale in China (Mainland). Customers from China (Mainland) please order the print book from: National Defense Industry Press.

This Springer imprint is published by the registered company Springer Nature Singapore Pte Ltd.
The registered company address is: 152 Beach Road, #21-01/04 Gateway East, Singapore 189721, Singapore

Preface

The design and fabrication of wind tunnel test models has a significant impact on the quality and efficiency of aircraft development. Due to the limitation of traditional processing methods, it is difficult to design and process models that must meet similar criteria of geometry, stiffness, mass, and dynamics, making it difficult for designers to evaluate the effects of aircraft structural parameters on aerodynamic characteristics in the early stages of aircraft development. Based on additive manufacturing technology (3D printing), new methods for designing and manufacturing wind tunnel test models have been developed significantly. The ability of additive manufacturing technology to print complex internal and external structures in an integrated manner can reduce the process constraints of manufacturing on design, increase the design freedom of wind tunnel test models, significantly reduce the number of parts of existing models, significantly increase the geometric similarity of shapes, and improve the model processing economy, which is more conducive to the development of new models with better similarity. This book introduces the common fundamentals of additive manufacturing technology based on wind tunnel test models, including design fundamentals, process fundamentals, inspection technology, and reinforcement technology. On this basis, the book introduces in detail the design and manufacturing technology of wind tunnel test models based on additive manufacturing with five types of wind tunnel test models, including conventional force measurement model, conventional pressure measurement model, elastic model, fluttering model, and deformation similar force measurement model and verifies the effectiveness of the new method by means of simulation calibration, ground test, and wind tunnel test.

The related research and the manuscript of this book were guided and assisted by academician Lu Bingheng from Xi'an Jiaotong University, Prof. Ding Xilun and Prof. Li Dongsheng from Beihang University. The participants of this research include Zhihua Zhou, Junhua Zeng, Xinglei Zhao, Wei Zhang, Kai Miao, etc. from Xi'an Jiaotong University and researcher Zhengyu Zhang, Dr. Yan Sun, Dr. Dangguo Yang, Dr. Chao Wang, etc. from the China Aerodynamics Research and Development Center.

The design and manufacture of wind tunnel test models is a multidisciplinary problem that requires continuous exploration and practice by researchers and engineers. There are many shortcomings and problems in the book, and we sincerely expect readers and experts to give criticism and correction.

Beijing, China Weijun Zhu
Xi'an, China Dichen Li

Introduction

This book introduces the common fundamentals related to additive manufacturing technology for wind tunnel test models, including design fundamentals, process fundamentals, inspection technology, and reinforcement technology. On this basis, the book introduces in detail the design and manufacturing technologies of additive manufacturing-based wind tunnel test models with five types of wind tunnel test models, including conventional force measurement model, conventional pressure measurement model, elastic model, fluttering model, and deformation similar force measurement model and verifies the effectiveness of the new method by means of simulation calibration, ground test, and wind tunnel test.

This book is mainly a reference for researchers and engineers engaged in the research of aircraft design, experimental fluid dynamics, and additive manufacturing technology.

Contents

Chapter 1
Overview of Wind Tunnel Test Models

1.1 Wind Tunnel Tests and Wind Tunnel Test Models

1.1.1 Aircraft Development and Wind Tunnel Tests

Aircraft development is a complex system engineering, including conceptual design, detailed design and trial production and qualification stages, of which conceptual design largely determines the performance of the aircraft and is the most important stage in aircraft development. It is the most important stage of aircraft development. In the conceptual design phase, the aircraft design includes aerodynamic design and structural design. The main task of aerodynamic design is to determine the aerodynamic shape parameters to minimize drag while ensuring sufficient lift, while the task of structural design is to select the structural layout form and determine the size of structural components to make the structure lightest in weight while ensuring structural integrity. In industrial practice, the relevant design work is carried out by the aerodynamic design department and the structural design department respectively.

The main means of aircraft development include simulation modeling (Modeling & Simulation), ground testing (Ground Testing) and flight testing (Flight Testing). As shown in Fig. 1.1 shows, according to the general process of aircraft design, flight testing is the closest to the real situation of the aircraft, and the data obtained have high credibility, but it can only be carried out at the later stage of aircraft design, after the prototype has been designed and manufactured, which has high cycle time and cost, and also has greater safety and other risks; with the improvement of computer computing power and the development of computing methods, CFD computing technology, MDO computing technology, etc. With the improvement of computer computing power and the development of calculation methods, CFD calculation technology, MDO calculation technology, etc. are expanding the range of problems that can be solved, and modeling simulation technology is increasingly used in the aircraft design process, but the reliability of the calculation is still

W. Zhu and D. Li, *Models for Wind Tunnel Tests Based on Additive Manufacturing Technology*, Additive Manufacturing Technology,
https://doi.org/10.1007/978-981-99-5877-1_1

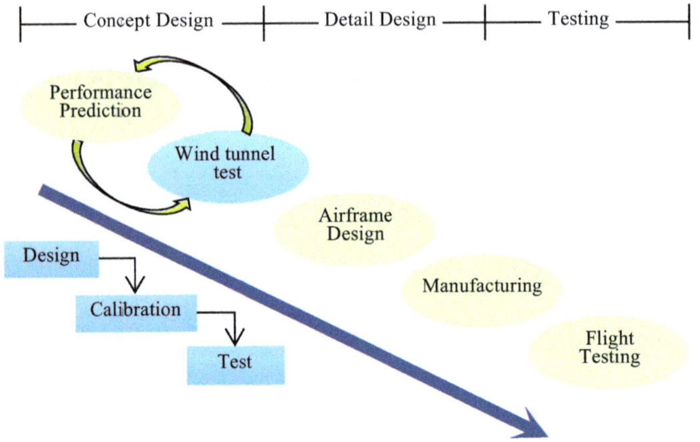

Fig. 1.1 The role of wind tunnel tests in aircraft development

waiting for test verification, and the accurate calculation of complex problems still requires a long calculation cycle and cost; ground test technology, mainly wind tunnel test, has a long history in the application of aircraft design, and has comparative advantages in terms of reliability and economy. Along with the development of modeling and simulation technology, the advancement of ground and flight test technology, and the improvement of data testing capability, the three technologies tend to converge in order to meet the new challenges in aircraft design [1]. Wind tunnel test data is the main basis of aircraft shape, strength and structure design and performance calculation, which is irreplaceable [1]. It is irreplaceable.

Wind tunnel tests are essential for aircraft aerodynamics and cross research fields with other disciplines, and their main roles include: (1) verifying the correctness of theoretical analysis and calculation in aerodynamics; (2) providing reliable aerodynamic data for the development of aircraft, missiles, rockets and other vehicles. The flight test uses real aircraft actually operating in a real environment, and the data obtained are the most reliable, essential for checking aircraft performance and testing flight envelopes in the late design stage, but there are test conditions that are difficult to control and change, unfavorable to the adjustment and measurement of test parameters, and the data obtained are scattered and difficult to be used to establish the intrinsic laws of physical phenomena. With the rapid development of computational fluid dynamics and the increasing capability of computer solutions, numerical computation can replace part of the wind tunnel test program and can provide aerodynamic data more rapidly and economically [2]. However, computational fluid dynamics is based on theoretical analysis and experimental hydrodynamics, which can solve the problem of clear physical mechanisms and sufficient experimental data, while its reliability is very challenging for the complex air flow phenomena frequently encountered in the development of advanced aircraft. Therefore, wind tunnel testing is still a means to balance reliability, safety and economy, and is widely used in aircraft type development and design research, finalization and modification, etc. It also keeps advancing the development of computational fluid dynamics.

1.1.2 Wind Tunnel Test Models

The wind tunnel test model is the test object of the wind tunnel test, and the physical model is designed and produced according to the similar theory, which is the substitute of the tested aircraft in the wind tunnel. There are various classifications of wind tunnel test models depending on the type of data acquired, similarity, model structure, test speed, etc., such as Table 1.1 shows.

The quality of the model design has an important impact on the reliability of the data and the safety of the wind tunnel tests. As shown in Fig. 1.2, the main requirements of model design include similarity requirements, testing requirements, strength requirements, and processability requirements. The similarity requirement refers to the similarity in the wind tunnel. The similarity requirement refers to the similarity between the model blowing in the wind tunnel and the aircraft flying in the atmosphere, including both object similarity and flow field similarity. The similarity of the flow field is determined by the blowing parameters of the wind tunnel and the aerodynamic shape of the model, while the similarity of the object means that the model design needs to meet similar conditions, including geometric similarity (similar shape), mass similarity, stiffness similarity, etc. as mentioned above.

The wind tunnel test model is based on the original aircraft (prototype) and designed according to the similarity principle. The common model similarity requirements include geometric similarity, stiffness similarity, mass similarity, and dynamic similarity. Geometric similarity means that the shape of the model is consistent with the prototype (after scaling, the same below) to ensure the similarity of winding, which is the basic criterion to be satisfied by all models; stiffness similarity requires that the structure of the model has the same stiffness distribution as the prototype,

Table 1.1 Types of wind tunnel test models

Classification Criteria	Wind tunnel test models
Get data type	Force measurement model, pressure measurement model, dynamic characteristic model
Similarity requirement	Geometrically similar, similar stiffness, similar mass, similar dynamics, similar structure
Model Integrity	Full machine model, half model, parts model
Test wind speed	Low-speed, subsonic, transonic, supersonic, hypersonic
Simulation object	Aircraft, missiles, rockets

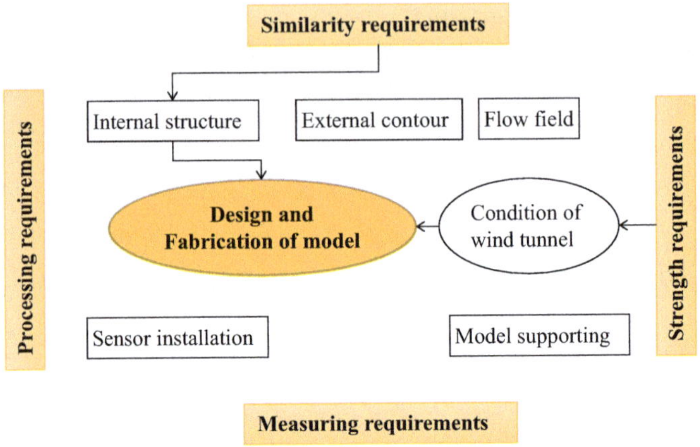

Fig. 1.2 Design requirements of wind tunnel test models [3]

which is required to be simulated for static elastic models, etc.; mass similarity requires that the model has the same mass distribution as the prototype, which is required to be satisfied for tailspin free-flying models, etc. Dynamic similarity is a requirement for the dynamic characteristics of the model, and models suitable for this criterion include fluttering models, buffeting models, etc. Structural similarity is a requirement for the similarity of the force transfer structure and path of the model, which is important for the accurate prediction of the effect of fluid–solid coupling.

1) Deficiencies in the similarity of traditional wind tunnel test models

(1) Inaccuracy of geometrically similar simulations

In order to solve the cost, for the aircraft shape of the detail features, generally do simplification, such as bulging; for the more complex parts of the structure, such as the fighter jet hanging bomb, often through the parts processing—overall assembly to achieve the way, which the integrity of the pneumatic shape will cause a certain impact. Detail simplification and assembly errors are problematic for the effectiveness of accurate aircraft design. Geometric scaling factor should be reasonably designed, generally requiring the model size to be as large as possible to increase the test Reynolds number, but in order to control the influence of cavity wall interference, the maximum size of the model (maximum cross-sectional area, overall length and wingspan length) should be kept within the specified range (depending on the type of test, wind tunnel size); at the same time, the surface roughness of the model should also meet the specified requirements;

(2) Inadequacy of structural parameter simulation

The structural parameters of the model are determined by similar stiffness, similar mass, similar dynamics and similar structure. As mentioned above, in the traditional wind tunnel test subjects, the introduction of such model tests is relatively late and rarely carried out, mostly "verification" tests (feasible or not), which is far from meeting the requirements for optimal design of structural parameters, and there is no way to carry out multidisciplinary optimization by means of tests. The similarity constraint of the existing vehicle scaled-down wind tunnel test model design is as follows:

a) Similar stiffness

Because of the complexity of the internal structure of the prototype aircraft and the reduced size of the model, it is more difficult to fully simulate the stiffness. The forms of model structures usually used are single/double beam-dimensional frame structure, beam frame structure, simplified similar structure, etc.

b) Similar quality

Mass similarity is usually achieved by using the structure mass-counterweight mass approach. For low-speed models, the structural mass is small and has a large counterweight adjustment space, so the mass similarity is easy to achieve; while for high-speed models, the structural mass itself is large in order to ensure sufficient strength and stiffness, and there is often an overweight problem, so the simulation of structural similarity is a difficult problem.

c) Similar power

The dynamic similarity has requirements for the stiffness distribution and mass distribution of the model, and involves more constraints; therefore, the accurate dynamic similarity design is more complicated, especially for the integral wing and full aircraft dynamic similarity models, which are very difficult to design and manufacture.

d) Similar structure

Structural similarity is a requirement for similarity of the model force transfer structure and path, which is important for accurate prediction of the effects of fluid–solid coupling effects. Theoretically, the model should have the same internal structure as the aircraft to accurately simulate the force transfer path. Theoretically, the model should have the same internal structure as the aircraft to accurately simulate the force transfer paths so that both have the same structural parameter properties. However, the complexity of the aircraft structure and thc scaled-down nature of the model make this economically unacceptable for conventional machining methods.

1.1.3 Design and Manufacturing Techniques for Existing Models

The main factor that causes the lack of similarity of the traditional models mentioned above is the model design and manufacturing technology—the wind tunnel test model implementation technology. The realization of the wind tunnel test model refers to the

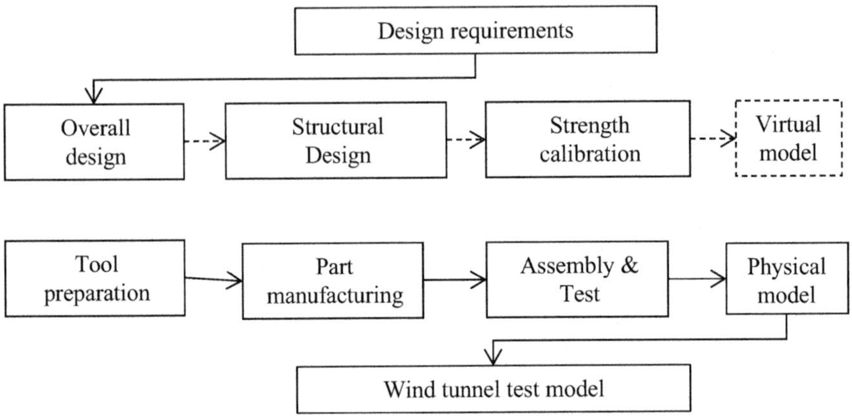

Fig. 1.3 Design and manufacturing process of wind tunnel test model

process of changing the model from demand to product, which specifically includes two parts: model design and model manufacturing. As shown in Fig. 1.3 the input of model realization is the model design requirement in the wind tunnel test task sheet, and the output is the model wind tunnel test. The broad model realization also includes modeling simulation, calibration calculation, calibration experiment and other links.

1) Product features of wind tunnel test model

Models are a special class of industrial products that have the following characteristics from the point of view of realization:

(1) Complexity of design constraints

As shown in Fig. 1.2, the design of the wind tunnel test model needs to meet similarity requirements, testing requirements, strength requirements and processability requirements, etc. There are many design constraints, and some requirements are contradictory to each other. For example, the static elastic model requires the model to have a certain stiffness distribution, but it is often difficult to meet the strength requirements for a model with similar stiffness when blowing wind at high speed; for the design of a dynamically similar model, the design with similar stiffness often brings overweight of mass.

(2) Single-piece form of production

During the development of an aircraft type, dozens of sets of models are designed and produced. Since the test purpose of each model set is different, the size and structure of the model are also different. Therefore, the processing of models is all single-piece processing, and the design and processing of special tooling for each set of models is costly and time-consuming.

(3) High punctuality of processing

Both military and civilian aircraft model development follow strict time node control, and all stages of model development must be completed within the specified time, so all aspects of model design, production and testing must be on time. Due to the multiple constraints of model design and single-piece model processing, it is necessary to go through complicated steps from the beginning of design to the completion of processing. Any mistake in the process will lead to the delay of model delivery and the postponement of wind tunnel test, which is unacceptable to both the model design units pursuing nodes and the test implementation units with compact test arrangements.

2) Model Design

The model design process includes the following aspects:

(1) Overall solution design

According to the model design requirements proposed in the wind tunnel test outline, the overall plan of model design is prepared, including the model size scaling factor, shape simulation requirements, model structure form, strength check conditions, material selection, processing accuracy requirements, support form and the size of scales and struts used, etc. This is the key step of model design.

(2) Structural Design

According to the proposed overall plan of the model, the detailed structure of the model is designed, and a full set of drawings of the model including general drawings, component drawings and parts drawings are drawn for processing.

(3) Strength calibration

The maximum aerodynamic loads are estimated based on the test conditions (Mach number, attitude angle, etc.) and the wind tunnel aerodynamic parameters of the model and its main components, and the strength of the model and the dangerous parts of its support rods are verified. Based on the strength check of the model, some adjustments are made to the structure and connections of the model and the materials used until the strength requirements are met.

3) Model processing

The materials often used in wind tunnel test models include metal (high performance steel and aluminum alloy), wood, glass fiber composite materials, etc., among which metal models are the main ones. The mechanical processing methods of metal models include wire cutting, EDM, CNC lathe, CNC milling machine, coordinate boring machine and other precision processing equipment. Tooling preparation is a necessary step for machining. It is an essential and complicated work for wind tunnel test model which has internal and external structure and high processing accuracy requirement. Aircraft fuselage and missile fuselage are generally made of ordinary steel or aluminum alloy; airfoil, model struts and balance are usually made of high quality high-strength alloy steel. At present, most of the metal models use CNC machining to manufacture the model components and leave a certain amount of post-processing machining allowance, and the whole assembly is manually customized and ground after completion. At the present stage of wind tunnel test model manufacturing, there is still a large part of the work to be completed manually, such as the flow line processing of the wind tunnel test model. The assembled model needs to be tested for accuracy of shape and surface quality before wind tunnel testing.

4) Features of existing implementation methods

The existing method has been developed for a long time and has a long technical accumulation. The design method and process, material selection, process preparation, and part processing are mature, and the parts realized by using high-precision machining technology have high dimensional accuracy and surface quality, and have an irreplaceable position in the realization of ultra-high-precision models (such as standard models). However, in view of the development trend of wind tunnel testing technology, the existing methods need improvement in the following aspects:

(1) Similarity design difficulties

In the premise of meeting the design requirements, the test model should be as light as possible, the current wind tunnel test model a large number of metal materials using CNC machining made of denser materials, weight reduction structure processing difficulties, so it is difficult to achieve model lightweight. For the tail spin free flight and other tests, the requirements to meet the quality of similar, its counterweight design and processing is more difficult. The internal structure of the aircraft is its main load-bearing component, which has an important influence on the structural parameters (stiffness distribution, mass distribution) and dynamic characteristics of the aircraft. However, the modulus of the model material is larger now, and the structural scaling of the model after design (e.g., 0.5 mm after scaling the skin with thickness of 10 mm by 1:20) is smaller than the machinable size, therefore, the structural similarity of the model cannot be achieved based on the present model implementation.

(2) Long delivery lead time

The shape and internal structure of the wind tunnel test model are complex, and the parts are generally produced in a single piece, with dozens of parts and hundreds of parts in 1 set of models. The process personnel have to design the processing plan according to the characteristics of each part, which is a great workload and time consuming [4]. According to the research, the delivery cycle of 1 set of full aircraft model is about 4 months, of which the process preparation takes about 1 month. There is a tendency to shorten the aircraft development cycle [1]. However, there is a limitation to shorten the cycle time of the model implementation, and the computational power such as CFD is improving, and the cycle time and cost advantages are becoming more and more obvious.

(3) Inadequate design approach

The design method of the present model is based on the mechanical processing method and widely uses CAD methods for model design, which is limited by the material selection, tooling preparation and other factors, and its design freedom is greatly restricted. It cannot take advantage of the latest design methods, such as structural optimization for weight reduction design, nor does it support well the integration with aircraft design methods, such as computational fluid dynamics (CFD) and multidisciplinary design for optimization (MDO), which does not coincide with the trend of wind tunnel test technology [1].

These shortcomings have hindered the development of new aircraft and prolonged the development cycle of aircraft. As the complexity of the model increases, the cycle time and cost of machining increases rapidly [5, 6]. The model's manufacturing cycle time or failure will delay the wind tunnel test. The extended manufacturing cycle or failure of the wind tunnel test model will delay the wind tunnel test time and make the scheduling of the wind tunnel test very difficult. This current situation requires the exploration of new manufacturing methods for wind tunnel test models.

1.2 Introduction of Additive Manufacturing Technology

WorldwideDomestic and foreign aerospace-related sectors are seeking a new manufacturing technology in order to reduce the time and cost of wind tunnel test model manufacturing [7, 8]. The main direction is the application of 3D Printing (or Additive Manufacturing-AM, or Rapid Prototyping-RP) technology in the manufacture of wind tunnel test models [9–12]. The main direction is the application of 3D Printing, or Additive Manufacturing-AM, or Rapid Prototyping-RP, technology to wind tunnel test models.

1.2.1 Introduction of Additive Manufacturing Technology

3D printing is a new manufacturing technology, different from the traditional mechanical processing "remove material" way, is a "add material" processing technology, is widely believed to have the potential to trigger a revolution in manufacturing [13]. The U.S. government has developed plans to use it as a core technology to revitalize U.S. manufacturing [14]. The U.S. government has developed plans to use this as a core technology to revitalize U.S. manufacturing.

3D printing technology is a processing technology that combines precision mechanics, material science, laser technology and CNC technology, and is known as a revolution in the field of manufacturing [15]. It is a manufacturing technology that uses the principle of material accumulation to make three-dimensional solids directly from virtual data. It is a manufacturing technology that uses the printing principle of

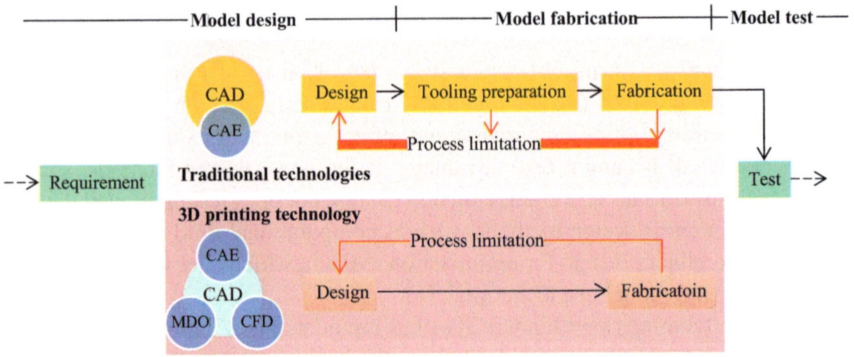

Fig. 1.4 Comparison of conventional machining and additive manufacturing technologies

material accumulation to make 3D entities directly from virtual data. Depending on the process, it can be divided into light curing process (Stereolithography, SL), fused deposition process (Fused Deposition Manufacturing, FDM), stacked manufacturing process (LOM), powder sintering process (SLS).

As shown in Fig. 1.4, the characteristics of 3D printing technology mainly lie in (1) rapidity: converting virtual parts (completed by CAD) into machining data (STL format, which can be converted automatically) and directly driving the equipment to produce solid parts, which can realize rapid manufacturing of products and is suitable for fields with strict requirements on product cycle time. (2) Highly flexible: eliminating the need for tooling preparation, the equipment can complete the processing and manufacturing of different types of parts without any changes or adjustments, which is suitable for new product development or single-piece small batch production. (3) Irrelevance to complexity: the manufacturing cycle and manufacturing cost of parts are not related to the shape and complexity of parts, but only to their net volume, which is suitable for the processing of products with internal and external complex structures. As shown, 3D printing technology greatly reduces the conversion distance from virtual parts to physical parts due to the elimination of tooling preparation, reaching a new level of seamless virtual-physical connection.

1.2.2 Application of Additive Manufacturing Technology in Wind Tunnel Test Models

A. Springer's team at NASA's Marshall Space Flight Center [5, 12, 16] investigated the feasibility of four 3D printed models made of different materials for use in a high-speed wind tunnel test model. In the comparison study, the aluminum model was used as the baseline for comparison, and the processes used for 3D printing were: fused deposition method (FDM); light-curing method (SL); laser-selected sintering method (SLS); and laminated layer method (LOM). A launch vehicle with a tail fin was used as the object of study, and the model was divided into two sections with a force balance installed inside, and the test Mach number ranged from 0.3 to 5. The tests showed that the SL model obtained satisfactory results under most test conditions; compared with the current design and manufacturing of metal models, the four 3D printed models reflected the characteristics of low cost and short design and manufacturing time. The use of 3D printing technology to fabricate wind tunnel test models can be used for early subsonic, sonic, and supersonic wind tunnel tests, especially the SL and FDM based low-speed rigid model fabrication technology has high accuracy, short cycle time, and low cost.

U.S. Air Force WPAFB Research Laboratory and Northrop Grumman Integrated Systems Research Group [17] used the 3D printing process to manufacture a fully resin-based rigid model of the Early Warning Aircraft (E-8C). Using the flexibility of 3D printing technology, the research team manufactured series models with/without radar fairing, with/without rudder deflection, and with/without slight winglet, while the good accuracy of the technology ensured the comparability between the series models. The study shows that the use of 3D printing technology to manufacture wind tunnel test models can improve the efficiency of wind tunnel experiments.

Boeing Corporation, USA [18] has identified rapid prototyping as one of two revolutionary technologies (the other being parametric CAD) to improve the efficiency and reduce the cost of wind tunnel testing. The lab has used metal SLS, Metal Coated Plastic, and Direct Metal Fusion to produce wind tunnel test models, and has shown that this process can reduce the cost of manufacturing models by an order of magnitude and triple the cycle time instead of traditional machining methods (NC programming, machining, hand grinding, etc.). The results show that this process can reduce the cost of manufacturing models by an order of magnitude and shorten the cycle time by three times, freeing designers from tedious model processing and even changing the way they work in the laboratory.

Jonathan D. Bartley-Cho et al. of Northrop Grumman, USA [19] used SL (Stereolithography system, SLS) to print the stage shell of a wing model and assembled it with a metal Al beam to form a dynamic aeroelastic model with a single beam-dimensional segmental structure to study the sudden wind mitigation performance (Gust Load Alleviation, GLA) of the wing structure, and obtained the expected results. The model is complex, but the main load-bearing element is an aluminum beam to simulate the stiffness; the aerodynamic shape is ensured by SL-printed segments, each segment is attached to the aluminum beam at a single point, and the gap between the segments is filled by foam; the mass similarity is achieved by installing lead blocks on the stage. Before the wind tunnel experiments, the team conducted a simulation check of the whole model, and the results showed that the static aeroelastic properties (static deformation) and flutter characteristics (flutter frequency and flutter speed) all meet the requirements.

G. Romeo of Politico Ditorino (Italy), P. Marzocca of Clarkson University (USA) and Illhan Tuzcu of University of Alabama (USA), among others, have used the SL process to fabricate a large span-to-skin ratio elastic wing shell for low-speed flutter experiments [20]. The SL shell is divided into 7 pieces with a small gap between them to reduce the additional stiffness of the shell, which guarantees a good aerodynamic profile accuracy.

In addition to this, (1) the U.S. Air Force WPAFB Research Laboratory and Johns Hopkins University's Applied Physics Laboratory [21, 22] (JHU/APL) collaborated to study the application of SL technology to wind tunnel pressurization model fabrication. The team designed, analyzed, and machined a low-cost LAMBD wing-body-bond wind tunnel pressurization model using the SL7000 former, and evaluated it with better experimental results. (2) NASA Langley Research Center [23] fabricated a wind tunnel test model of hypersonic aerospace spacecraft by 3D printing-based precision casting technology. They use light-curing 3D printing technology (SL) to manufacture monolithic small-size aerospace wind tunnel test models, which can reduce the weight of the model and improve the design speed and reduce costs. (3) The U.S. Air Force WPAFB Research Laboratory and Dayton University have jointly conducted research work on additive manufacturing technology for wind tunnel test models [24, 25]. The group successfully fabricated two wind tunnel test models by 3D printing technologies (SL and SLS): using low-cost 3D printing technology to process the outer surface of the model and regular metal parts for the internal strength support parts; the twisting and deformation of the wing should be minimized; the pressure measurement holes and pressure transfer pipes can be effectively processed and electronic or mechanical pressure scanning valves and sensors can be installed relatively easily; the support and the model should be able to detect the head angle accuracy and precision under loading. (4) Heyes et al. of Imperial University London, UK [7] discussed methods and techniques for manufacturing wind tunnel test models with complex internal structures using fused filament deposition manufacturing (FDM) and light-cured 3D printing (SL). It was found that the light-curing 3D printing technology not only shortens the model fabrication cycle and reduces the cost, but also enables the rapid fabrication of pressure measurement flow paths and models with complex internal structures while ensuring the requirements of stiffness and strength. (5) D. B. Landrum et al.of Alabama University (USA) [26] specifically investigated the feasibility of light-curing forming in the fabrication of aerospace wing models. The group fabricated two wing models for evaluation, using polyurethane through a common casting technique and resin light-curing molding and sanding the surface with water-based sandpaper. The team concluded that the light-curing 3D printing technology provides better aerodynamic surface accuracy. (6) The Central Aerodynamic Research Institute of Russia and the Russian Academy of Sciences collaborated on the application of light-curing forming and its composite manufacturing technology in the design and manufacture of fighter models [27]. They studied the feasibility of using light-curing 3D printing to manufacture higher precision wind tunnel test models. The preliminary study proved that among the 3D printing methods, light-curing 3D printing is the most effective means of manufacturing aerodynamic models, which can produce individual model parts as

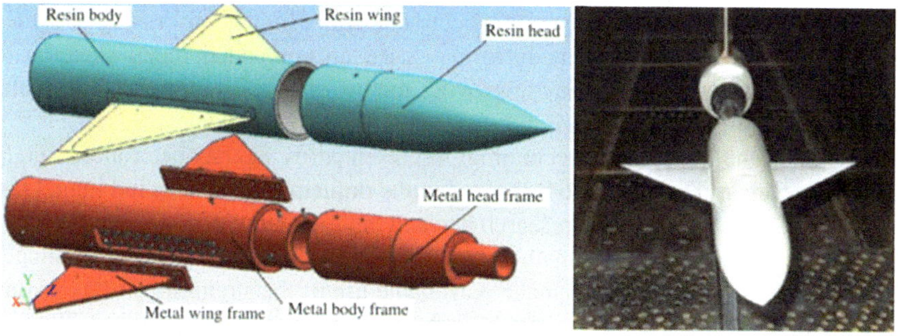

Fig. 1.5 Compounded AGARD-B force measurement model [29–31]

well as composite manufacturing of combined parts and models with high precision. (7) The Department of Mechanical Engineering, McGill University, Canada, studied the effectiveness and economy of 3D printing technology in the manufacture of wind tunnel test models [11]. The study found that 3D printing technology is particularly suitable for the fabrication of non-structural load components, and classified the fabrication techniques of wind tunnel test model components into three categories: non-structural load, light load and high load components. (8) Aghanajafi et al. at K. N. Toosi University of Technology, Iran [28] fabricated a low-speed, transonic wind tunnel test model using 3D printing technology and compared with a CNC machined aluminum alloy model. At the same time, the feasibility of the 3D printing manufacturing technology is evaluated to meet the high fidelity shape requirements.

The design and manufacturing of wind tunnel test models based on light-curing 3D printing technology has a more mature basis, and some preliminary explorations and applications have been carried out [29–31]. As shown in Fig. 1.5, Xi'an Jiaotong University, China has made some attempts in the manufacture of force measurement model, the design and manufacture of pressure measurement holes and flow channels of pressure measurement model, and the manufacture of composite wind tunnel test model by electrodeposition, and has also achieved certain research results. However, further research work is needed to push the technology to model applications.

1.2.3 Analysis of the Current Status of the Application of Additive Manufacturing Technology

Light-curing 3D printing (Stereo-Lithography, SL for short) is currently recognized as one of the most widely used among the many 3D printing methods [7, 12, 21, 30, 32–35]. Compared to other 3D printing processes (FDM, LOM, SLS, etc.), light-curing 3D printing is currently recognized as the process with the highest printing accuracy (0.1 mm for conventional devices and up to 0.05 mm under special conditions). possibility. Recent studies have shown that SL, on the other hand, proves to have better application prospects due to its higher processing accuracy [12, 15, 32]. Some researchers have also used an embedded metal skeleton to strengthen SL components, thus improving the bending resistance of wind tunnel test models in high-speed blowing tests [19, 31, 36].

Although the 3D printing technology represented by SL has been increasingly used, there are still problems to be solved in the implementation of wind tunnel test models as follows. The above research has laid the foundation for the establishment of new model implementation methods, but in order to give full play to the advantages of 3D printing technology and make reasonable use of the strengths of the existing model implementation methods, there is a need to continue research in the following aspects.

1) The influence of processing technology on design methods

Nowadays, most of the research on the implementation of wind tunnel test models based on 3D printing technology is limited to the discussion at the level of model machining, while the influence of another aspect of model implementation, model design, is less studied. As a constraint (machining processability and machining economy), the way the part is machined must have an impact on the way the part is designed and its efficiency. The removal of tooling preparation and the realization of seamless virtual-solid conversion not only shorten the cycle time significantly in the sense of machining, but also help to expand the choice of model design tools and improve the efficiency of model design by reducing constraints. Thus, the introduction of 3D printing technology affects both model design and model machining, thus having an impact on the overall model realization approach.

2) Integration with traditional mechanical processing methods

The advantages of 3D printing technology are the shortening of the conversion cycle from virtual parts to solid parts, the rapid processing capability of complex structures, etc. However, it is not yet able to replace traditional mechanical machining methods in the realization of high precision and high strength models. Therefore, the respective advantages of 3D printing technology and mechanical machining methods can be integrated, and their respective ranges of application can be fully considered in the model design, and the corresponding machining methods can be selected for different structures of the model, thus providing a new comprehensive solution for the realization of wind tunnel test models. In addition, most of the existing studies are feasibility studies, and most of the selected objects are simplified wing structures, which generally do not include details such as control surfaces and rudder surface declination adjustment mechanisms. These are important for the practicalization of new manufacturing methods for wind tunnel test models, and need to be studied specifically. Therefore, it is necessary to make full use of the respective advantages of 3D printing technology and mechanical processing methods for more productized prototypes, and expand more detailed technical aspects to establish the foundation for the new design of the model and the practicalization of the manufacturing method.

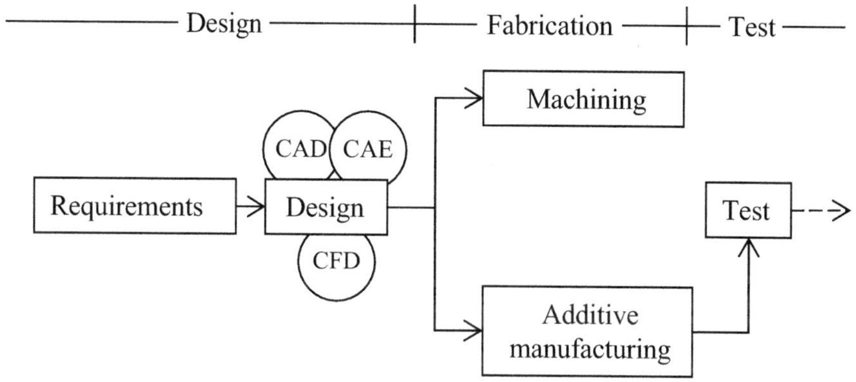

Fig. 1.6 SL-based wind tunnel test model implementation method

Using the mainstream 3D printing technology—SL process, model design and model manufacturing as a whole, this book proposes a new design and manufacturing method for aircraft wind tunnel test models, and the principle of the method is as shown in Fig. 1.6.

This new model realization method expands the tooling options for model design, combines the advantages of traditional mechanical processing methods and SL technology methods, and utilizes the performance characteristics of polymer materials used in the SL process, which can improve the similarity of wind tunnel test models, improve the realization efficiency of existing model types, and provide support for the development of new aircraft development technologies.

3) Rethinking the properties of polymers

As mentioned above, polymer materials are commonly used in current 3D printing processing technology. According to the current model design and processing standards, polymer materials are not included in the recommended catalog. There is a view that polymer materials are not suitable for wind tunnel test model manufacturing, and the main reasons are the higher damping coefficient, lower elastic modulus, lower material strength and poor time stability of polymer materials. If the traditional metal model is used as a reference, the polymer model has a natural deficiency and its performance cannot reach the "standard value". However, if one thinks differently, the perception of high-index materials may be different. The low modulus of polymer provides a possibility to solve the structural similarity that is difficult to achieve with existing metallic materials; the higher damping coefficient helps to improve the stability of rigid models, and for dynamic models, the high damping characteristics of the material itself can be balanced by reducing the total number of parts with the advantage of 3D printing integrated processing, thus reducing the structural damping; the lack of strength of polymer materials can be balanced by embedded metal The lack of strength of polymer materials can be compensated by embedded metal, surface metal plating, etc. Therefore, based on such new understanding, it is possible to design and manufacture new wind tunnel test models and even develop new wind tunnel test techniques.

1.3 Overview of Additive Manufacturing Technology for Wind Tunnel Test Models

1.3.1 A New Framework of Design and Fabrication for Wind Tunnel Test Models

Based on additive manufacturing technology, the new framework for the design and manufacturing of wind tunnel test models is shown in Fig. 1.7, including requirement analysis, model design and manufacturing (printing), model calibration and wind tunnel test. In the model design segment, according to the test requirements and similarity requirements, designers obtain the corresponding parameters of the model from the parameters of the vehicle prototype. In particular, due to the introduction of additive manufacturing technology to increase the model design freedom, more automated design methods can be adopted, such as CAD and CAE fusion for optimal design methods. Once the design results (e.g. CAD) have been processed and the process parameters have been set, the 3D printer can manufacture the model accordingly. To ensure the accuracy and safety of the tests, the model must pass strength calibration, resonance calibration, stiffness calibration, vibration calibration, etc. Finally, the model is installed in a wind tunnel for testing to obtain aerodynamic data.

Design and manufacturing of models. The design and manufacturing process is shown in Fig. 1.8. The whole process is divided into two phases, "virtual model" and "physical model" [37]. The inputs are various design requirements, and the outputs include wind tunnel test models and test outlines, etc.

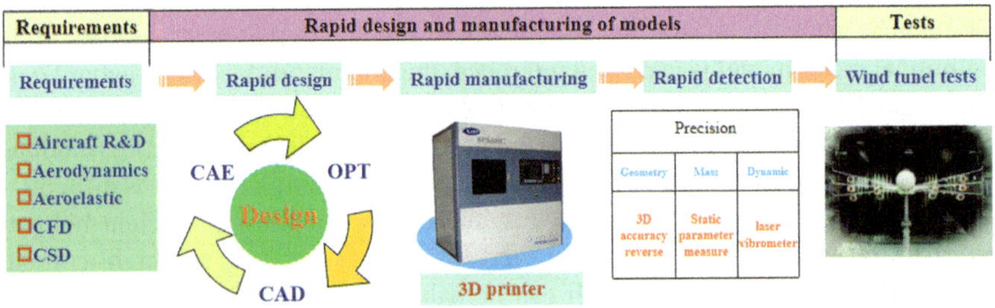

Fig. 1.7 A new framework of design and fabrication for wind tunnel test models

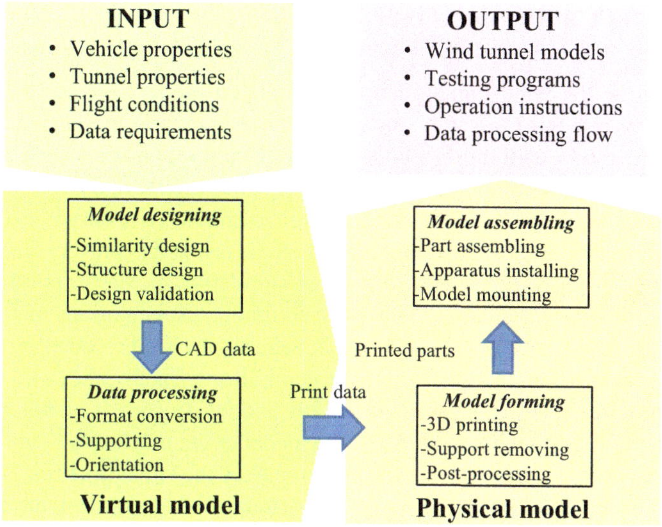

Fig. 1.8 Flow chart of the design and manufacturing of the models

Model design: The design of wind tunnel model starts from similarity design, i.e., the corresponding parameters of the model are obtained from the geometric and structural parameters of the prototype of the vehicle according to the similarity principle. Under the constraints of similarity requirements, testing requirements, process requirements, etc., the model design results are obtained by CAD and other tools [38]. The model design results are obtained by CAD and other tools under the constraints of similarity requirements, testing requirements and process requirements. After this result completes the design verification [39]. The CAD data of the model can be obtained.

Data processing: Data processing is a special part of 3D printing processing, which is used to generate the database that drives the operation of 3D printers. Generally, the common formats of AM data include STL, AMF, etc. Therefore, CAD data needs to be converted to the above data formats. Nowadays, the mainstream CAD design software provides data processing interfaces. Part support design and print direction design are necessary parts of AM process preparation [40].

Model forming: Once the data is available, model forming is done automatically by the 3D printer. The formed parts need to be removed from the supports, cleaned of residual raw materials, etc. To improve the material properties and surface roughness, the model can be post-cured (e.g., light-cured [40]), sandblasting treatment, etc. [37] ….

Model installation: Thanks to the molding capability of 3D printing technology, the structural integrity of the model is improved and the number of parts is significantly reduced [41]. This allows for a simplified model assembly. During the model assembly process, sensors and devices for data acquisition, etc. are mounted to the model. After fixing the final model to the wind tunnel support (usually a balance [42]. The design and machining process of the model is completed.

1.4 Model Calibration Overview

Before starting the wind tunnel test, the completed model needs to be calibrated in order to ensure the test safety and verify the accuracy of the design. Generally speaking, the model calibration includes strength calibration, resonance calibration, ground stiffness calibration test, ground modal calibration test, etc.

Strength calibration: The purpose of strength calibration is to ensure that the model will not be structurally damaged in the wind tunnel test and to prevent damage to the experimenter or the wind tunnel. Numerical simulation methods are often used to verify the strength of a model. First, the wind tunnel test conditions that will cause the maximum load need to be determined. Second, the loads loaded on the model are calculated based on the wind tunnel test conditions. Finally, the loads are loaded into the model and the stress distribution of the model structure is obtained using the FEA method. If the maximum stress of the model at the maximum load is less than the required stress of the material, the design passes the strength check.

Resonance calibration: In order to avoid resonance of the model during the test, which may cause damage to the model, wind tunnel and even personnel, the resonance frequency of the model-balance-support system should be designed to be higher than the peak frequency of the wind tunnel caused by the flow field. This frequency is determined by its stiffness, damping and mass properties. Since the first two factors depend mainly on the unchangeable balance and support, the main design variable is the weight of the model. Typically, the lower the weight of the model, the higher the system resonance frequency [43]. 3D printing supports the use of lightweight structures and materials, which can reduce the weight of the model and have a positive impact on improving the safety of wind tunnel tests.

Stiffness calibration: For models that need to meet the stiffness similarity conditions (such as static aeroelastic models, dynamically similar models, etc.), the assembled models need to be ground-checked to determine whether the stiffness distribution of the model meets expectations [44]. The principle of this test is to load a specific force at a specific point and invert the stiffness distribution of the model by measuring the deformation of the model [43]. The measurement of the deformation of the model is critical. Accurate measurement of the model deformation is the key, and non-contact measurement methods such as optics can be used [45]. Non-contact measurement methods such as optical can be used.

Vibration calibration: The purpose of ground vibration testing is to calibrate the dynamic characteristics of the model (such as resonant frequency and resonant mode), etc., for dynamic model design and manufacturing accuracy [19]. In a typical test, the model is first fixed on a special fixture. In a typical test, the model is first fixed on a special fixture, and under the excitation (shaker, etc.), the model's response in the time domain is obtained by sensors or non-contact testing means, and then the dynamic characteristics parameters of the model can be obtained after data processing using Fast Fourier. In a typical test, different dynamic parameters are obtained by different excitation methods [46]. The non-contact vibration test method reduces the number of the non-contact vibration test method is widely used for vibration calibration of models because it can reduce the interference of sensors and improve the testing efficiency [44]. The non-contact vibration test method is widely used for vibration calibration of models.

1.5 Typical Equipments and Materials

Most of the wind tunnel test models in this book use the SPS600B light-curing 3D printer of Xi'an Jiaotong University, as Fig. 1.9. The printing range is 600 mm × 600 mm × 450 mm. Its printing range is 600 mm × 600 mm × 450 mm, and the printing accuracy is ±0.1 mm. The typical printing process parameters are shown in Table 1.2.

The materials used in this book include two categories of resins and metals, and their mechanical properties are as shown in Table 1.3. The resin grade is SOMOS ®14120 (DSM, Netherlands), and the material preparation is done by SPS600B light-curing printer (Xi'an Jiaotong University).

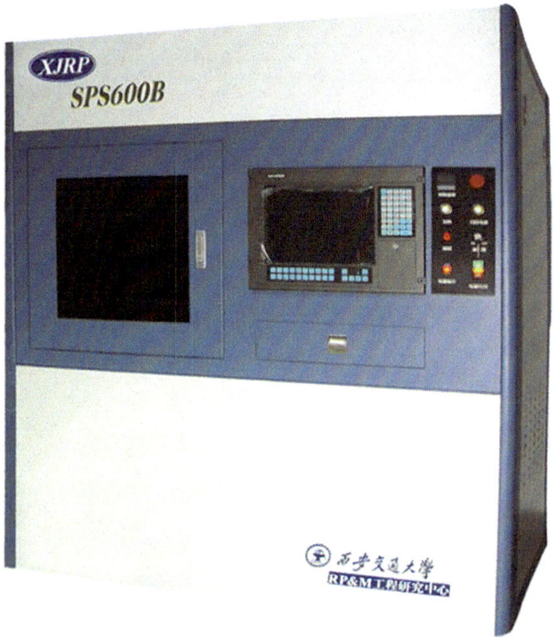

Fig. 1.9 SPS600B light-curing 3D printer

Table 1.2 Main printing parameters for model manufacturing

Parameter category	Parameter name	Parameter values
Basic parameters	Delamination thickness/mm	0.1
Scanning parameters	Filling spacing/mm	0.1
	Laser power/mW	198.2
	Spot compensation diameter/mm	0.12
	Fill scan speed/mm s^{-1}	3600
	Supported scanning speed/mm s^{-1}	1200
	Point support scan time/ms	1.66
	Hinge structure scan time/ms	1.37
	Contour scanning speed/mm s^{-1}	3500
	Jump speed/mm s^{-1}	12,000

Table 1.3 Performance parameters of the materials used in the model

Parameters	Resin SOMOS 14120	Steel 40Cr	Steel 45	Aluminum alloy 7A04
Density, ρ/kg m^{-3}	1120	7780	7850	2780
Young's modulus, E/GPa	2.46	211	211	74.0
Poisson's ratio, ν	0.38	0.30	0.30	0.30
Tensile strength, σ/MPa	46.0	980	600	600

Chapter 2
Design Basis of Wind Tunnel Test Models Based on Additive Manufacturing

In order to ensure manufacutirng effectiveness, manufacturing-oriented design of the wind tunnel test models is required, including shape adjustment design, Active manipulation surface design, segmentation and connection design, and so on.

2.1 Wind Tunnel Test Model Shape Adjustment Design

2.1.1 Model Shape Adjustment

1) Plugging cone design

Generally, in order to obtain accurate experimental data, the model should simulate the flow of aircraft inlet and tail nozzle [2]. However, this test model does not use the ventilated air inlet design. If the air inlet and tail nozzle are simply blocked, the result will definitely make the airflow characteristics of the model not similar to those of the actual aircraft. When the aircraft is in flight, the airflow around the air inlet and tail nozzle of the aircraft is as Fig. 2.1a shows that part of the incoming flow into the intake, and then ejected from the tail nozzle, the rest of the uniform flow around the outside, usually without airflow separation; if the simple inlet and tail nozzle blocked, the flow characteristics as Fig. 2.1b shows, this time around the lip of the inlet and near the tail nozzle to produce an obvious airflow separation, around the model of the airflow and the aircraft is very different. Therefore, in order to better simulate the airflow around the aircraft, a streamlined rotating body is usually designed in the inlet and tail nozzle of the model, i.e., a rectifier block.

W. Zhu and D. Li, *Models for Wind Tunnel Tests Based on Additive Manufacturing Technology*, Additive Manufacturing Technology,
https://doi.org/10.1007/978-981-99-5877-1_2

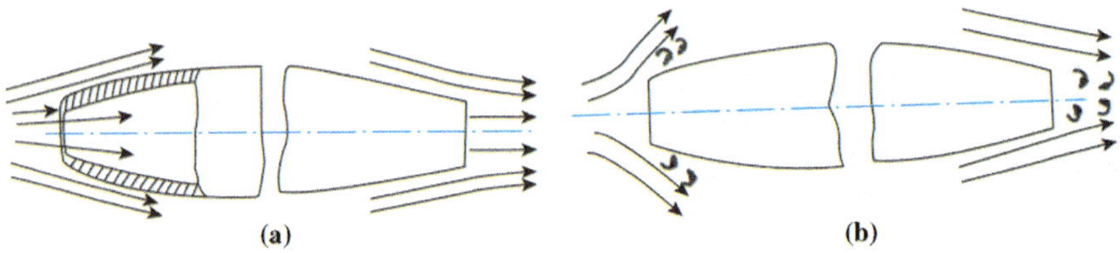

Fig. 2.1 Simulation of airplane model inlet and tail nozzle. **a** Aircraft bypass simulation, **b** simulation of simple blockage of model inlet and tail nozzle

According to the design requirement of [31], the model adopts parabolic rectifier block, which is a rotating body with parabolic shape. The length-to-thin ratio of the block is 4, and the half cone angle of the tip of the block is 15°, while the shape of the block is consistent with the lip of the air inlet, and it is smoothly connected with the airframe, as shown in Fig. 2.2.

For the tail nozzle plug can be used symmetrical shape contour line. The tail nozzle plug of a model uses a straight line segment profile extending 30 mm outward, as shown in Fig. 2.3.

Fig. 2.2 Aircraft model air inlet block design

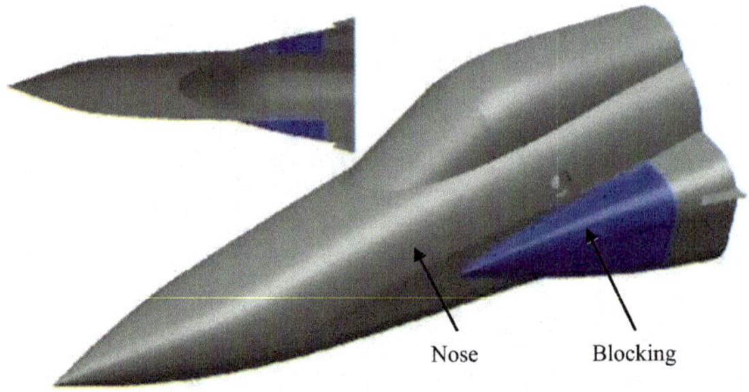

Fig. 2.3 Model aircraft tail nozzle plug

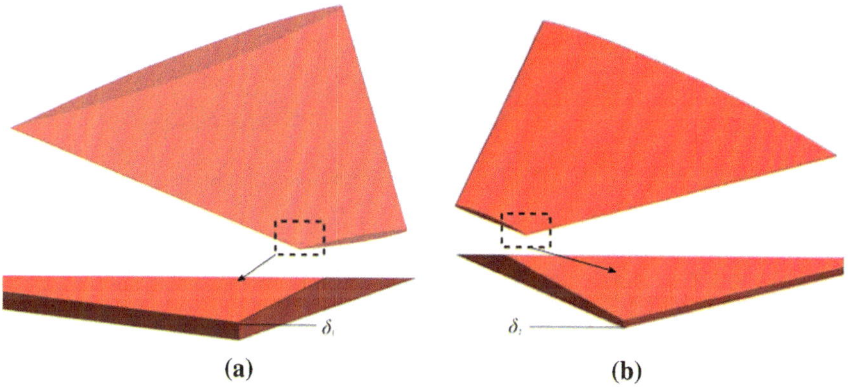

Fig. 2.4 Model thin edge thickness. **a** Trailing edge of flat tail, **b** trailing edge of wing

2) Local enlargement design

In the model design, the trailing edge of the manipulation surface of the model is a thin edge of very small thickness. As shown in Fig. 2.4 the thickness of the trailing edge of the flat tail $\delta_1 = 0.07$ mm and the thickness of the trailing edge of the rear flap $\delta_2 = 0.09$ mm. The structure of this thin edge is difficult to realize in actual machining. For processing convenience, the thin edge of the trailing edge of the maneuvering surface needs to be designed with partial thickening.

At present, the processing accuracy of 3D printing is $\pm$ 0.1 mm, i.e., the thickness of the processed layering is 0.05–0.1 mm. If the thin edge size of the manipulation surface of the model is less than 0.1 mm, the thin edge is easily incomplete or missing when printing, which affects the shape dimensional accuracy of the manipulation surface of the model, and then affects the aerodynamic characteristics of the model.

The thin edge of the manipulation surface is partially thickened. As shown in Fig. 2.5, in order not to affect the overall shape of the manipulation surface, the offset section is appropriate at the last 2–3 mm of the trailing edge, and the trailing edge is offset by 0.1 mm outward along the center line, so that the thickness of the thin edge of the trailing edge is greater than 0.2 mm.

3) Baseline marking design

After the aircraft model is assembled, in order to measure and adjust the position accuracy and angle of the model, it is necessary to make engraving lines on the model, the following are the engraving line requirements for the model [31]:

(1) The horizontal reference line of the fuselage;
(2) The line of intersection of the fuselage reference plane (i.e. the fuselage symmetry plane) with the fuselage;
(3) The intersection of the wing chord plane with the wing, the leading and lateral edges of the other airfoils;

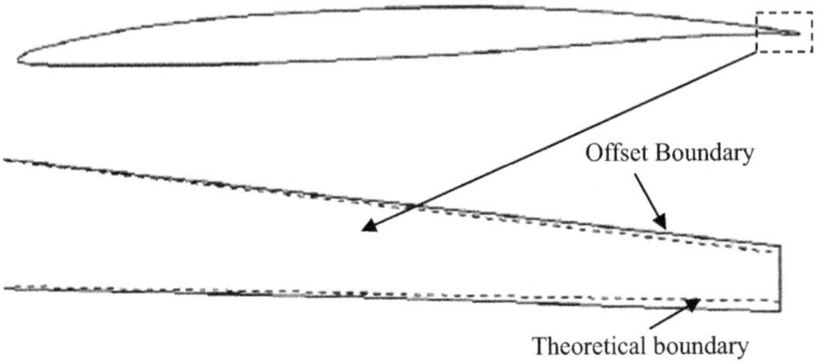

Fig. 2.5 Thickened design of the trailing edge of the maneuvering surface

(4) The centerline of the external object and its position line on the wing and fuselage;
(5) The width of the scoring line is less than 0.2 mm and the depth is less than 0.2 mm.

The engraving of metal models needs to be done after the processing of the model shape is completed, which is difficult and takes a long time to process. Light-curing 3D printing has the processing advantage of printing any complex structure at one time, so we can consider processing and printing the model shape parts and model engraving at the same time, such as Fig. 2.6.

2.2 Active Manipulation Surface Design for Wind Tunnel Test Model

2.2.1 Overview of Model Manipulation Surface Design

According to the test outline of the test, if the efficiency of each rudder surface is to be measured, it is necessary to ensure that the flaps, ailerons, rudder, elevator or flat tail on the model can be deflected within the required range, and the relative position of each maneuvering surface rotation axis should be the same as that of the prototype. In order to achieve the rudder surface deflection, some small parts of the rudder surface maneuvering mechanism of the model that do not simulate the prototype are allowed to be exposed to the airflow. The rudder surface manipulation mechanism should be convenient and accurate when changing the rudder surface deflection angle, and it should ensure that the deflection angle of each manipulation surface does not change due to the aerodynamic force during the test. The technical requirements of the manipulating surface of the wind tunnel test model include [2]:

(1) Accurate shape: Maneuvering surfaces are designed in accordance with geometric similarity requirements, and certain small parts that do not mimic the prototype cannot be exposed to the airflow;

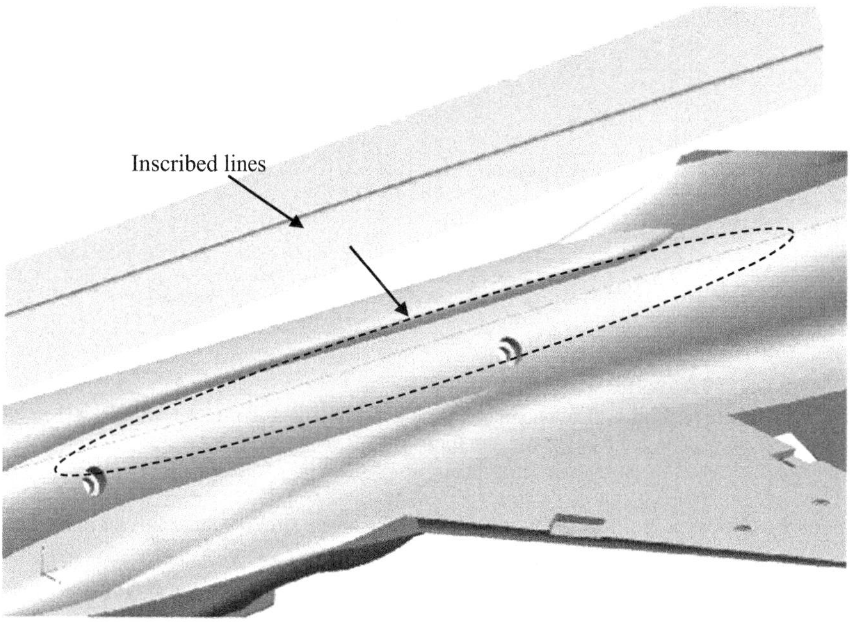

Fig. 2.6 Aircraft reference line design

(2) Adjustable angle: the manipulation surface can be deflected within a certain angle range, and the deflection angle for the manipulation surface should be precise;
(3) Structural reliability: the connection structure should have sufficient rigidity and strength to ensure that the manipulation surface declination does not change during the experiment due to the action of aerodynamic forces;
(4) Easy to disassemble: connection structure should be easy to disassemble, for repeated use of the structure, should also consider the durability of the mating surface.

Traditional metal models generally use two forms of pivot pin positioning and variable angle piece positioning. Figure 2.7a is the rudder deflection method, because there is not enough structural space to install and fix the rotor shaft, only different rudder models with different angles can be made, and during the experiment, the rudder deflection angle can be changed by installing different rudder models, also this deflection method can be applied to the front flap, aileron, rear flap and other structures; Fig. 2.7b is the structure method of deflecting the flat tail, through the pin hole; Fig. 2.7b is the structure of deflecting flat tail, which can make a pair of flat tail meet different angles of flat tail deflection through the pin holes. The design of the maneuvering surface of the resin-metal composite model also refers to Fig. 2.7. However, it is necessary to optimize the design of these two solutions to ensure that the connection strength of the resin material manipulation surface structure can

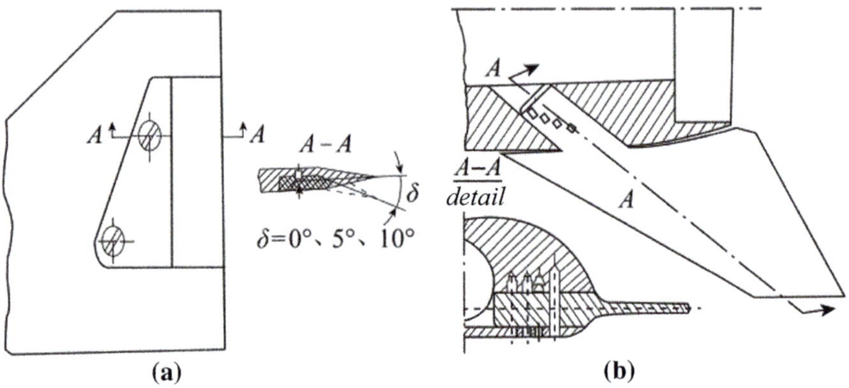

Fig. 2.7 Structural form of manipulation surface of wind tunnel test model. **a** Deflected rudder structure form, **b** deflected flat tail structure form

meet the experimental requirements, and at the same time, to take advantage of the processing advantage of 3D printing without the complexity of the model, whether the model and its components that need to be divided when processing the metal model but are actually unnecessary can be printed as a whole by 3D printing method.

For larger models, the manipulation surface can be positioned by means of angle plates, positioning plates, positioning pins and clamping hinges, as shown in Figs. 2.8, 2.9, 2.10 and 2.11. The advantages and disadvantages of various connection positioning methods are compared in Table 2.1. Among them, the variable angle piece is a common positioning method, in which a series of angle pieces are machined according to the test content and the rudder declination is changed by replacing different angle pieces. If the angle needs to be increased during the test, a new angle piece is simply machined. The use of angle pieces to achieve different rudder declination is more time-consuming, and changing the rudder declination is also more troublesome. However, this way the model surface is clean, the connection is reliable, and the repeatability of angle positioning is high.

Fig. 2.8 Rudder declination by angle piece. 1—Airfoil; 2—Angle piece; 3—Maneuvering surface

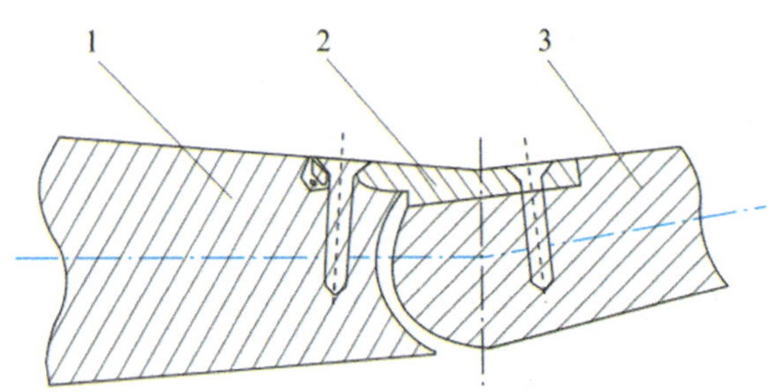

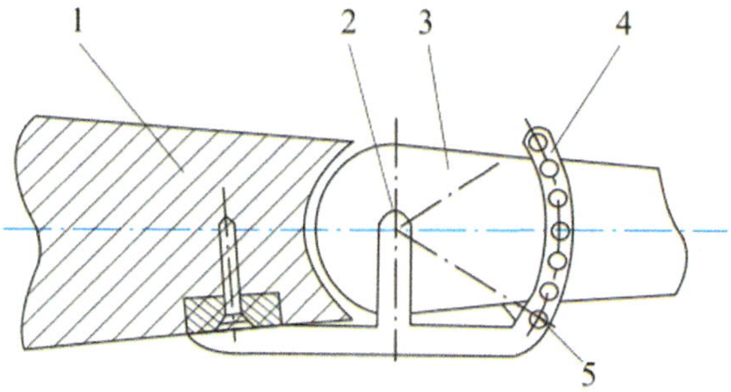

Fig. 2.9 Rudder declination with positioning plate. 1—Airfoil; 2—Rotary shaft; 3—Manipulation surface; 4—Locating plate; 5—Locating pin

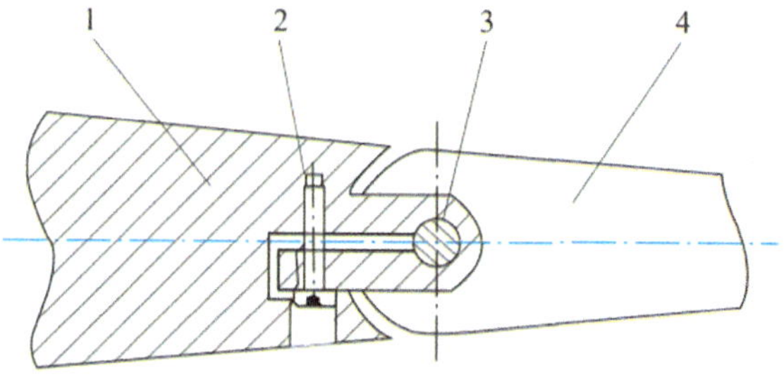

Fig. 2.10 Rudder declination with clamping hinge. 1—Airfoil; 2—Clamping screw; 3—Swivel; 4—Maneuvering surface

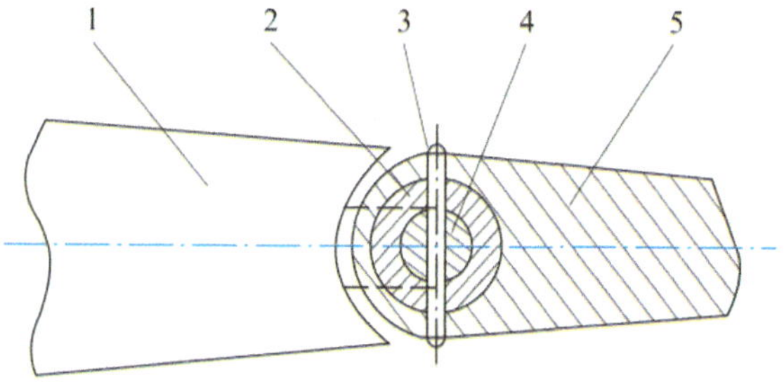

Fig. 2.11 Rudder declination by means of bushings. 1—Airfoil; 2—Bushing; 3—Locating pin; 4—Swivel; 5—Maneuvering surface

Table 2.1 Positioning method of rudder deflection angle

Rudder deflection mode	Advantages	Disadvantages
Variable angle piece	Clean model surface and reliable connection High repeatability of angular positioning	Model processing is more time-consuming Changing the declination is more troublesome
Positioning plate	More accurate positioning and convenient angle change	Interference with airflow
Clamping hinge (generally not recommended)	Continuous airfoil processing	Must angle sample Prone to looseness and structural damage
Positioning pins	Convenient and reliable	More complex to manufacture

For smaller models, it is difficult to use various rudder surface steering mechanisms mentioned above because of the small size of the steering surface and the small aerodynamic load on the steering surface. In this case, the transition fit of the shaft hole can be used to achieve the rudder surface deflection angle adjustment. Using this approach, angle samples should be provided.

2.2.2 Suitable for 3D Printing of Variable Angle Piece Design

In the model manufacturing method based on light-curing printing, the model and its components do not have the problem of structural complexity, and the model and its components that need to be divided during CNC machining but are actually unnecessary can be printed as a whole by the 3D printing method. Also, considering the size of the model and its components and the size of the load bearing, the following two options are designed:

1) Integrated printing of variable angle sheets and manipulation surfaces

As Fig. 2.12 shown, in the small-sized maneuvering surface components, the maneuvering surface does not have enough thickness to install the metal vario-angle piece, so the solution of integrating the vario-angle piece with the maneuvering surface is adopted. The pure resin maneuvering surface structure reduces the assembly link and has high assembly accuracy, but in the test, several angles of the maneuvering surface need to be tested, several angles of the resin maneuvering surface have to be manufactured, so it cannot be replaced by one resin maneuvering surface, and therefore, it cannot be replaced by a resin manipulating surface, and the strength of the connection between the manipulating surface and the main wing is low and cannot bear excessive load.

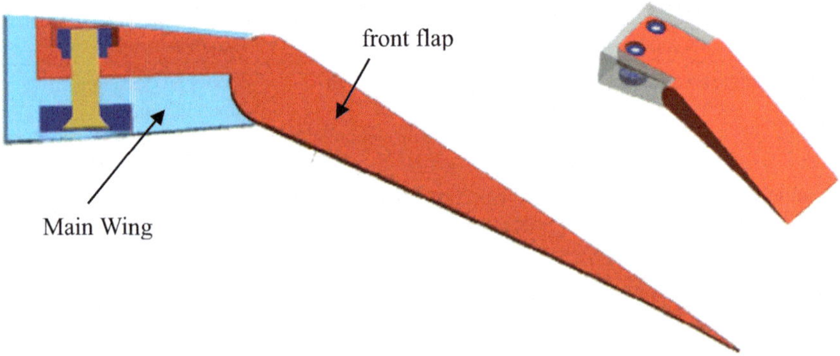

Fig. 2.12 Variable-angle sheet and flap or aileron integrated manufacturing

2) Separate printing of the angle change piece and the manipulation surface

As Fig. 2.13a shows, in the large size manipulating surface parts, the manipulating surface has enough thickness to install the metal varicaps, and the deflection of the manipulating surface can be realized by installing varicaps with different angles, so only one resin manipulating surface can meet the test requirements of multiple deflection angles, and the resin manipulating surface fixed on the varicaps has high connection strength, which can improve the connection strength of the manipulating surface. However, the CNC milled metal angle change piece cannot fully satisfy the pneumatic shape of the model, and generally the shape has to be repaired with putty. Meanwhile, since the secondary assembly is used, the manufacturing accuracy of the metal angle change piece and resin manipulating surface needs to be strictly controlled to ensure the assembly accuracy of the manipulating surface.

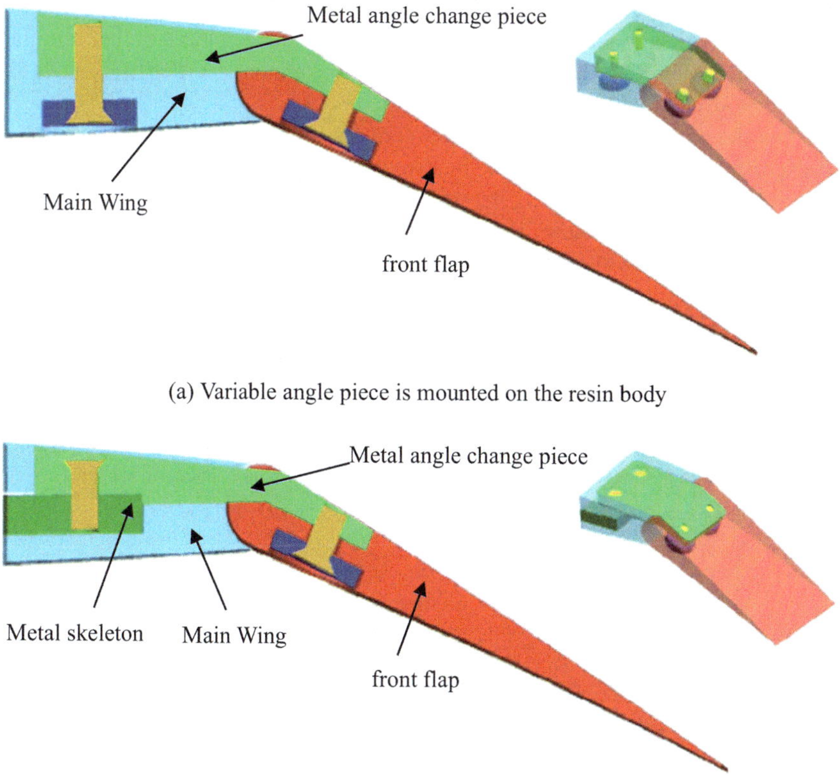

(a) Variable angle piece is mounted on the resin body

(b) Variable angle piece is mounted on the metal skeleton

Fig. 2.13 Variable angle sheet printed separately from flaps or ailerons

As Fig. 2.13b shows, the metal skeleton can be extended to the part of the metal angle change piece in the large size model design. By attaching the metal angle change piece to the metal skeleton, it can maximize the connection strength of the maneuvering surface and ensure that the maneuvering surface can maintain the accurate deflection angle even under the large aerodynamic load.

Analyzing the above two design solutions, we can see the advantages and disadvantages of each of the two solutions, as shown in Table 2.2. The model size is small and the experimental load is small, so the variable angle piece and the manipulation surface can be manufactured integrally; the model size is larger, the manipulation surface changes many angles, and the experimental load is large, so the variable angle piece and the manipulation surface can be printed separately.

Combined with the structural size characteristics of this model, the front flap, rear flap, aileron and rudder of the model are printed separately from the maneuvering surface, as shown in Fig. 2.14. If the maneuvering surface is large, more than one angle-change piece can be used to fix the connection. The front flap of this model is fixed with two angle-change pieces, and the angle-change pieces of the front flap, rear flap and aileron are fixed on the metal wing skeleton in order to improve the connection strength of the maneuvering surface.

When the deflection axis of the maneuvering surface is not on the center line of the chord surface at both ends of the maneuvering surface, the maneuvering surface will produce position deflection after deflection, and there will be a large gap or even interference when connecting with the main wing, so different angles of deflection need to be realized with different angles of the front flap parts. For example, in this model, since the axis of deflection is on the lower surface, two sets of circular transition front flaps are designed in the two sets of rudder surface deflection to reduce the gap in the assembly. Table 2.3 shows the composition of the maneuvering surface parts for the applicable variable angle sheet form.

Table 2.2 Advantages and disadvantages of the two flap and aileron connection schemes

Flap aileron connection structure	Advantages	Disadvantages
Variable angle sheeting integrated with flaps or ailerons	Pure resin First-class assembly with high precision	Costly resin, high cost Low connection strength
Variable angle sheet printed separately from flaps or ailerons	High connection strength Save resin material	Secondary assembly, affecting accuracy Time consuming processing of metal angle change pieces and inconvenient flow line repair with resin surface
Variable angle piece mounted on metal skeleton		

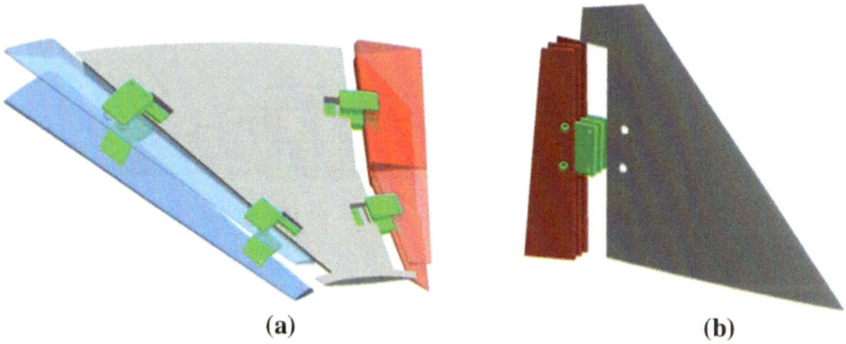

Fig. 2.14 The effect of the manipulation surface design of the variable angle piece. **a** Wing maneuvering surface design, **b** drogue rudder design

Table 2.3 Composition of manipulating surface parts in the form of variable angle pieces

	Front flap	Backlap	Aileron	Steering rudder
Resin	Left 0°, 24° Right 0°, 24°	Left and right 1 set	Left and right 1 set	1 set
Variable angle piece	Left 0°, 24° Right 0°, 24°	Left 0°, 25° Right 0°, 25°	Left 0°, − 15° to 25° Right 0°, 15°25°	0°, 15°, 25°

2.2.3 *Suitable for 3D Printing of Pivot Pins Design*

Taking advantage of the processing advantages of light-curing 3D printing, the shape and manipulation surface parts of the model are made of resin; the main deflection positioning and load-bearing parts, such as the fixed base, and the rotary axis are made of metal. The base is fixed and connected in the resin profile, and the rotary axis is fixed and connected in the resin manipulation surface, and the base and rotary axis are fixed by pin holes when assembled, as shown in Fig. 2.15.

The base, as the main load-bearing component, is installed and fixed in a way that determines the reliability of the entire deflection structure, and the following two installation methods are mainly adopted:

(1) Base and metal skeleton connection

This solution is easy to position and simple to install, and because the base is connected to the metal skeleton, the whole metal skeleton acts as a force-bearing, and its connection strength is high to ensure that the maneuvering surface deflection angle does not change during the test due to the aerodynamic force. However, this also increases the design difficulty of the metal skeleton and the installation difficulty of the resin housing components.

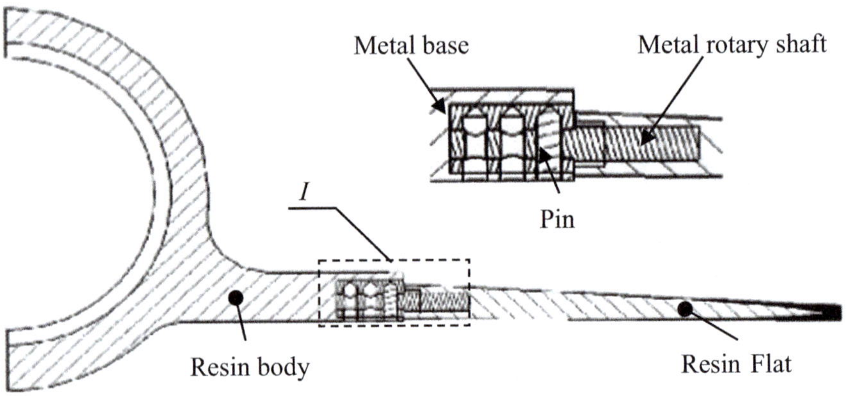

Fig. 2.15 Rotating pin structure

(2) Base embedded in resin parts and fixed with adhesive

The connection strength of this scheme is the design key. In order to increase the bonding performance between resin and metal, the base should be designed as large as possible, try to increase the size of L, B and H, increase the contact area between metal and metal, and roll the grid around the surface of the base to increase the mosaic effect of the colloid and the surface to enhance the bonding performance. Also ensure that the wall thickness δ of the resin is above 5 mm to avoid thin wall thickness and rupture under larger load, as Fig. 2.16 shown.

The rotating shaft parts are also embedded in the maneuvering surface and then fixed with adhesive. The flat surface of the chord like the flat tail makes it difficult to have space to insert the rotating shaft into a deeper position, and the volume and area occupied by the rotating shaft is small compared to the flat tail, so it cannot bear too much load, as shown in Fig. 2.17.

The flat tail part of this model is far away from the tail skeleton, and it is difficult to design the connection between the base and the skeleton, so the base is embedded in the resin part and then fixed with adhesive. As Fig. 2.18 and Table 2.4 shows the composition of the maneuvering surface parts for the variable angle piece form.

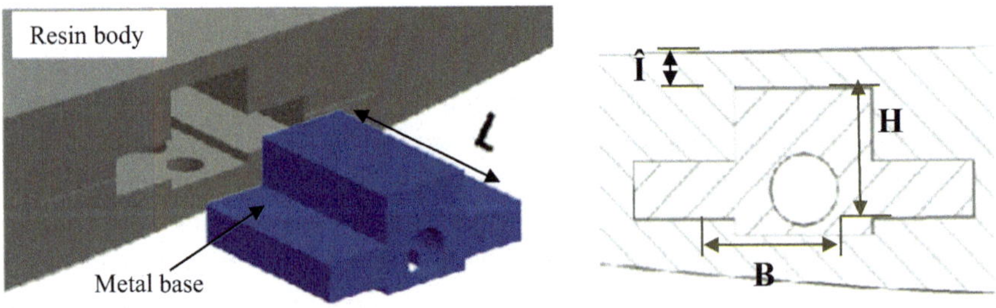

Fig. 2.16 Metal base design. **a** Resin body with metal base, **b** cross-sectional view

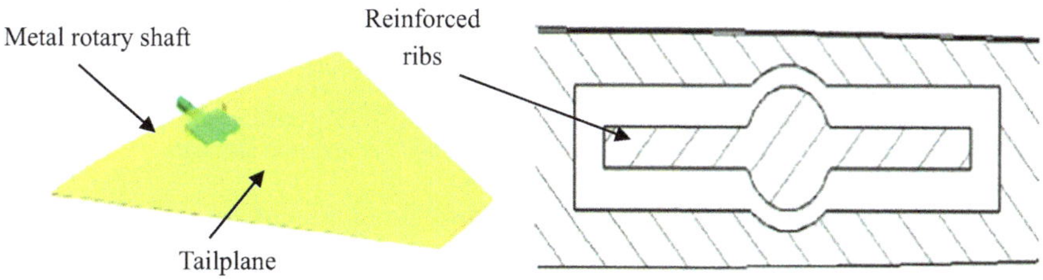

Fig. 2.17 Metal rotary shaft design. **a** Resin flat tail with metal swivel shaft, **b** cross-sectional view

Fig. 2.18 Flat tail design effect

Table 2.4 Composition of manipulating surface parts in the form of rotating pins

	Hirao
Resin	Left and right 1 set
Metal	Left and right metal bases, left and right metal swivel shafts

2.2.4 3D Printed Rudder Surface Connection Design

In CNC machining, for the convenience of manufacturing, the wings, flaps and angle change pieces are often processed separately, and the deflection of the maneuvering surface is achieved by replacing the angle change pieces with different deflection angles. In the 3D printing-oriented manufacturing mode, there is no problem of the complexity of the model, and the model that needs to be divided during CNC machining but is not necessary can be printed as a whole by the 3D printing method. If required, the variable angle piece can be manufactured in conjunction with the flap. The comparative analysis of the advantages and disadvantages of various assembly connection methods is shown in Table 2.5.

Table 2.5 Comparison of the advantages and disadvantages of the flap connection scheme

<table>
<tr><th colspan="2">Flap connection structure</th><th>Advantages</th><th colspan="2">Disadvantages</th></tr>
<tr><td colspan="2">Integrated manufacturing of variable-angle flaps and flaps (for small size parts)</td><td>Pure resin; first-class assembly, high precision</td><td colspan="2">Costly resin, high cost, low connection strength</td></tr>
<tr><td rowspan="3">Variable angle flaps and flaps individual printing (for large size parts)</td><td>Pure resin</td><td>Easy and convenient</td><td rowspan="3">Save resin materials</td><td>Low intensity</td></tr>
<tr><td>Metal Processing</td><td>High connection strength and high precision</td><td>Time-consuming processing and inconvenient flow line repair with resin surface</td></tr>
<tr><td>Metal-resin composite</td><td>Good accuracy and high strength of metal positioning surface, reusable</td><td>Manufacturing complexity</td></tr>
</table>

(1) Integrated manufacturing of variable-angle flaps and flaps: Applicable to small-sized parts, a large number of interchangeable flaps need to be manufactured, and each flap corresponds to a certain deflection angle, thus using a large amount of resin, but there is only one level of assembly, and the assembly accuracy is relatively high.
(2) Variable angle piece and flaps are printed separately: Applicable to large size parts, the position angle of flaps is adjusted by each variable angle piece with different declination, secondary assembly, which affects the assembly accuracy, but saves the use of resin material. It can be divided into three forms: resin, resin-metal composite, and metal.

Pure resin angle change piece completely printed by resin, can be cured sleeve with screws and angle change piece assembly hole interference fit, angle change piece need to leave a 0.5 mm fit margin, apply external force to destroy the size of the angle change piece aperture spin assembly, this method angle change piece can not be reused, should be prepared more than one angle piece to meet the use of demand. The traditional metal processing method has good processing accuracy and connection strength, but the processing is time-consuming and does not facilitate the flow line with the resin profile surface repair.

The design of the resin-metal composite angle change piece is divided into two parts, as shown in Fig. 2.19. The part that fits to the chord plane is made of metal with regular shape design, which is convenient for processing; the part that fits to the airfoil of the aircraft is processed by resin 3D printing method, and the resin shape is easy to be resharpened and convenient to match the streamline with the airfoil. The metal and resin are bonded to form the whole of the variable angle piece, and the bonding strength can be increased by changing the roughness of the bonding surface, and the bonding positioning of the upper and lower parts is completed by using the way of setting through-holes for the metal and tabs for the resin.

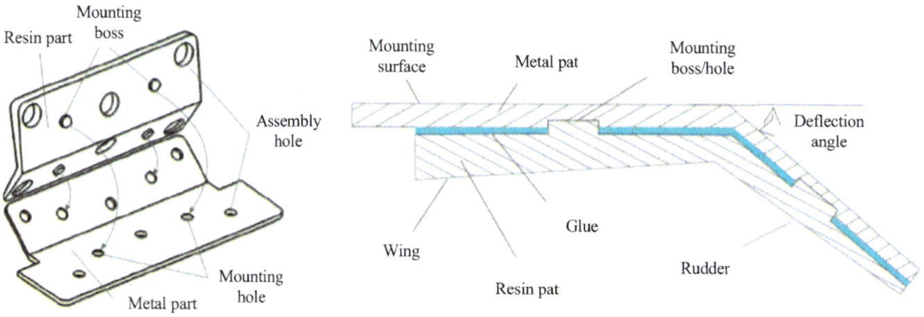

Fig. 2.19 Resin-metal composite angle change piece. **a** Schematic diagram of angle-change piece connection and positioning, **b** composite angle-change piece cross-section

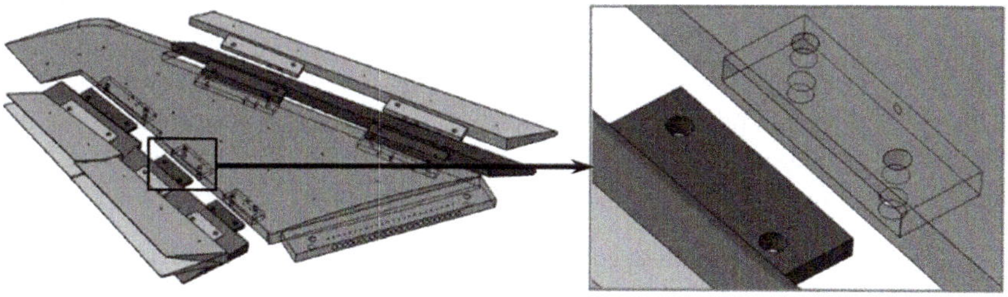

Fig. 2.20 Built-in angular deflection method

The above method uses the traditional form of varus sheet, which is placed only on one side of the wing chord plane, and part of the airfoil is scribed into the varus sheet, which will have some effect on the profile surface. Since 3D printing does not have this trouble of machining deep grooves in the thin part of the wing, this book also considered the method. The assembly structure shown in Fig. 2.20, where the variable angle piece is placed in the middle of the chord plane to keep the wing profile continuous and intact, forming a slotted connection, and the assembly connection is made with pins, which also requires metal reinforcement of the connection part.

2.2.5 *Reinforcement Method for Resin Assembly Parts*

Due to the brittleness of resin material and the disadvantage of easy wear, metal parts can be inlaid in the assembly part to overcome the brittleness of resin, improve the connection strength and ensure the reuse of each part without damaging the connection surface. The metal insert is used for the connection of screws, etc., so it is also necessary to ensure the coaxiality of the hole during assembly to avoid serious tilting or misalignment during the threaded connection, so the following two methods of positioning the assembly hole are proposed.

(1) Auxiliary positioning method

As shown in Fig. 2.21, the axis of the connection hole is perpendicular to the chord plane, so the chord plane is used as the axial positioning surface of the sleeve and the template, and the mounting boundary surface of the variable angle piece assists the hole position of the template to determine the radial position of the sleeve. The positioning template is laminated to the chord plane of the resin model and the installation boundary surface of the variable angle piece, and the positioning holes of the template are CNC machined, and the pins are installed and the sleeve is set to determine the radial direction of the connecting sleeve. The maximum length of the sleeve should not be higher than the aircraft airfoil where the bonding hole is located to prevent the impact on the airfoil shape, and then insert the sleeve on the pin to prevent the flow of the bonding agent into the sleeve countersink hole. The outer surface of the sleeve is rolled mesh to increase the mosaic effect between the adhesive and the surface to enhance the bonding performance; the gap between the outer surface of the sleeve and the bonding hole is taken as 1–3 mm, in order to facilitate the injection flow of viscous resin or sufficient UV irradiation of photosensitive resin, while also ensuring the bonding strength, the connection gap can be adjusted according to the height of the bonding hole.

(2) Direct positioning method

As shown in Fig. 2.22 the 3D printed model's own stepped hole surface is used as the positioning reference, and the metal sleeve is directly pressed into the positioning hole after applying adhesive, thus leaving a certain gap at the non-positioning surface to facilitate the curing of the adhesive connection.

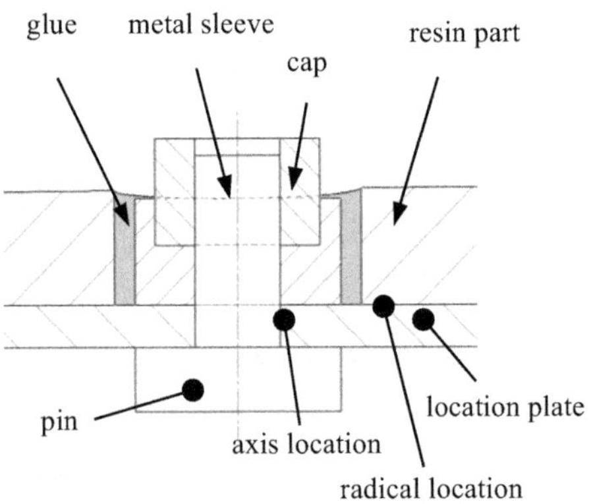

Fig. 2.21 Schematic diagram of auxiliary positioning

Fig. 2.22 Schematic diagram of direct positioning

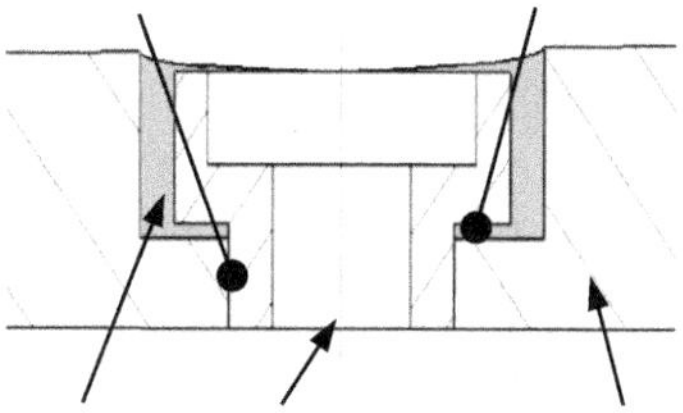

2.3 Segmentation and Connection Design of Wind Tunnel Test Model

Wind tunnel test models are developing in the direction of large-scale and monolithic, while the processing size range of 3D printing equipment is limited (e.g., the maximum processing size of SPS600B light-curing 3D printer of Xi'an Jiaotong University is 600 mm × 600 mm × 450 mm). For some larger model parts, 3D printers cannot finish processing at one time, which requires dividing the large parts into several smaller sub-parts to be processed separately and then finally bonded or assembled into a whole. On the other hand, due to the different shape characteristics of the model parts, it is sometimes difficult to ensure that all parts of the parts are printed in one direction with high precision or form a more uniform error. If the parts are reasonably divided, the printing process can be optimized separately according to the need, thus improving the overall manufacturing accuracy of the model.

2.3.1 Wind Tunnel Test Model of Splitting Design

When the model is designed, it should be combined with the test content specified in the test syllabus and the processing range of the 3D printer, and the model should be split appropriately. The reasonable splitting of the model makes the parts process well, reduces the manufacturing cost and improves the manufacturing accuracy; meanwhile, it makes the parts position accurately, connect reliably and disassemble easily.

1) **Principle of division**

The segmentation of the wind tunnel test model is firstly oriented to the segmentation of the test requirements. When designing the model, the model should be decomposed into parts according to the test content specified in the test outline. Decomposition of parts requires reasonable, easy disassembly; parts connected to reliable, accurate positioning; parts of good craftsmanship, low manufacturing costs; parts interchangeability is good, using different forms of assembly to meet the various requirements of the test. For example, in the aircraft selection test, the model can be used to measure the aerodynamic characteristics of the whole aircraft, but also requires the measurement of the contribution of each component to the aerodynamics. Therefore, each part of the model (such as wing, fuselage, tail, hair room, landing gear, external bracket, etc.) should be easily disassembled and assembled. When doing the test of individual parts, it is required that each part itself can maintain a smooth aerodynamic shape; when doing the test of the whole aircraft, it can be easily installed and the repeatable positioning accuracy of the parts can be ensured; when conducting the experiment without the wing, the wing can be removed and a wing plug block can be installed to keep the surface of the fuselage with a smooth shape.

In determining the structure of the model, it should also consider the support mode of the model in the wind tunnel, and the requirements of the model and wind tunnel equipment. For example, try to arrange the installation position and space of the scanning valve in the inner cavity of the pressure measurement model, and the pressure measurement tube, cable and pipeline pulled out from the model should be led from the position with little bypass interference. For some special test models, such as the power motor and water-cooled pipeline with power test, the layout of power cables, special measuring instruments (hot-wire anemometer, acceleration sensor, actual angle sensor, etc.) installation coordination, blowing or jet test of high-pressure gas pipeline, etc., in the model design are required to leave a suitable space or reasonable laying. Therefore, in order to avoid the influence of the instrument on the aircraft blowing flow field, and to facilitate the installation of various sensors and piping system laying, the aircraft wind tunnel test model also needs to be processed in separate parts.

The wind tunnel test model should reduce the number of divisions to reduce the impact of assembly errors. A typical set of model structures is roughly divided into the following parts:

(1) The fuselage part: is the model of the load-bearing parts, the fuselage has the main force members, due to the slender fuselage, to use the segmentation structure, according to the needs of the general from 2 to 4 sections.
(2) Airfoil part: It is the force-bearing part of the model, including the wing, horizontal tail, vertical tail, and bullet wing, etc.
(3) Movable maneuvering surface part: including wing leading and trailing edge flaps, ailerons, elevator, rudder, angle piece, hinge mechanism, etc.
(4) Power and ventilation lines: including lip devices, ventilation lines y, engine compartment, nozzle and nozzle, etc.
(5) Connection and support device: mainly with the wind tunnel connected to the support rod, the sky and other components.

2) **The number of splits and the choice of split location**

According to the test requirements of the wind tunnel test model, accuracy requirements, etc., combined with the working range of the printer to determine the number of sub-model divisions, the overall layout of the chunk. Since the assembly of sub-models inevitably brings a certain degree of accuracy loss, and this error may increase a little with each increase in the number of sub-models, unnecessary partitioning should be avoided as much as possible to keep the number of sub-models as small as possible.

When the printing parameters are certain, the deformation of the sub-model is mainly determined by the structure and manufacturing direction of the sub-model, so the manufacturing direction of the sub-model should be considered when the specific partition is made. For large parts segmentation manufacturing, in order to ensure the same amount of deformation on both sides of the same segmentation line and the same deformation direction, two adjacent sub-model manufacturing should try to ensure that the spatial direction of the segmentation section is the same, in order to reduce the overall deformation after assembly. In order to reduce the deformation of the sub-model, the manufacturing direction should also try to make the sub-model structure with small deformation or Z-direction deformation-based easy-to-control cross-section when printing, such as priority use of non-slender cross-section, annular cross-section, etc. The size of the split section should be moderate, not too large, as far as possible in the shape of the more regular or less strict size requirements, otherwise the flatness of the plane will be difficult to ensure, is not conducive to the assembly and bonding of the sub-model; also not too small, try to avoid cutting at the thin wall, otherwise the positioning structure will have nowhere to add, but also affect the assembly strength of the parts.

The deformation of the model manufacturing is related to the size of the model, so the size distribution of the sub-model is also considered in the specific division. As shown in Fig. 2.23, the deformation of sub-model 1 along the common division line is different from the deformation of sub-models 2 and 3 along the line, and the deformation of sub-models 2 and 3 along the other division line is also different, so it is difficult to splice smoothly on this division line in the end. Therefore, the size of the sub-models should be the same or similar when splitting, so as to avoid the splicing situation of less to more on one split line.

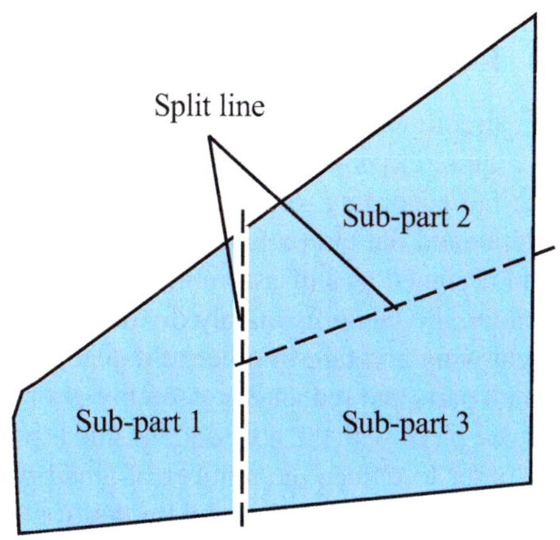

Fig. 2.23 Less-to-many splicing on a split line

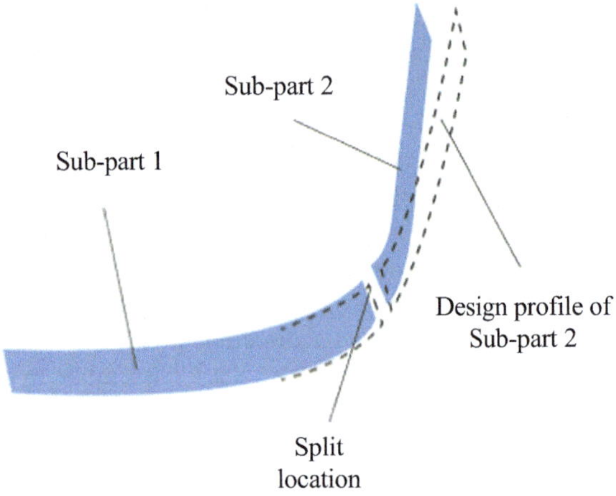

Fig. 2.24 Large deformation due to local curvature change

Surface-based model, due to the sub-model manufacturing deformation not only Z-direction warpage, there is also deformation within the XOY surface, this deformation is generally expressed as a reduction in the curvature of the arc, so that it tends to be straight, in the case of a small difference in arc length, the curvature of the arc deformation is larger. The specific model of the cross-section type is generally a combination of arcs of different curvature, the local deformation on both sides of the split line has a great impact on the assembly and the deformation after assembly, such as Fig. 2.24. In order to reduce the deformation after assembly, it can be seen that the change of local curvature may cause a large deformation after assembly. In order to smoothly assemble and reduce the overall deformation after assembly, the partition line should be chosen at the curvature of the surface as small as possible. The surface of the wind tunnel test model is curved, and the streamline surface is not complicated and has few abrupt changes, so it can be divided appropriately according to the curvature change.

3) **Example of segmentation design**

An aircraft test model with fuselage length L = 1420 mm and wingspan b = 850 mm is required to conduct efficiency tests of front flaps, ailerons, rear flaps, rudder and flat tail according to the test syllabus. Therefore, the front flap, aileron, rear flap, rudder and flat tail parts of the model should be separated from the fuselage model and designed as a disassembled structure; according to the processing range of 3D printer, the model is mainly divided into five parts: nose, front fuselage, left wing, right wing and tail; in order to reduce the separate manufacturing and assembly of small parts and reduce the assembly error of the model, the air inlet block is printed as one piece with the nose and the gun is printed as one piece with the front fuselage. In order to reduce the number of small parts, the air inlet plug is printed with the nose, the cannon is printed with the front fuselage, and the tail, ventral fin and tail are printed in one system. As shown in Fig. 2.25 the resin parts of the model were finally split into 14 parts.

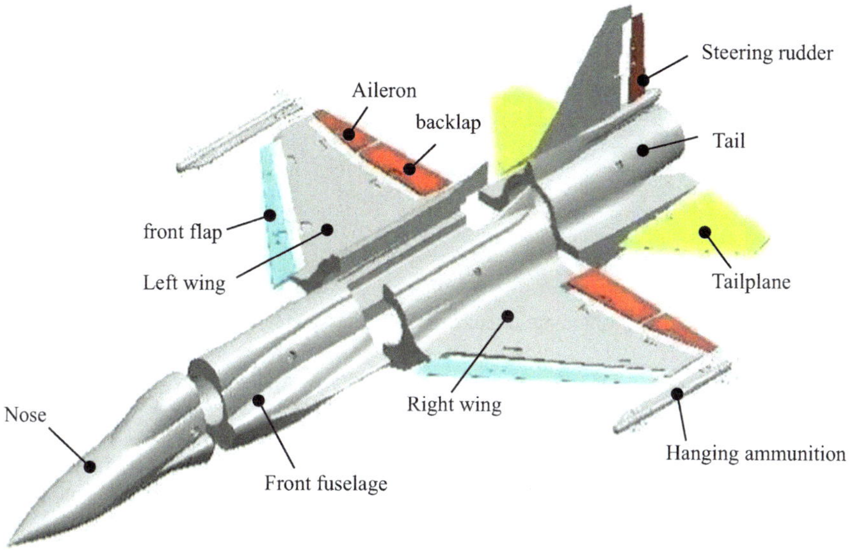

Fig. 2.25 Segmentation design scheme of a wind tunnel test model of a certain type of aircraft

4) **Design of assembly auxiliary structure**

The division should be considered in a good assembly way. Unlike the test-oriented assembly, the model division considered here is no longer required to be removable after assembly. The adhesive is directly used for bonding, and a certain bonding gap is also cut out between the model assembly sections. At the same time, the model is required to meet the strength of the connection after assembly, and a mounting slot can be designed in the part that allows redesign, and the metal reinforcement plate can be embedded in the mounting slot and bonded, and a certain bonding gap should be reserved between the reinforcement plate and the mounting slot. In addition, the assembly also needs positioning structures, such as tabs, recessed holes, etc., through the positioning structure to assist the model of accurate assembly. The positioning structure is often square or round, and in order to ensure better positioning and to meet the strength requirements after bonding, it can be cut as shown in Fig. 2.26. The size and number of positioning steps on the partition line are related to the size of the sub-model and the length of the partition line, but the number of positioning steps on a partition line should not be too many, 2–4 is appropriate, too many may cause assembly difficulties. In addition, due to the impact of 3D printer processing accuracy, the best between the tab and the concave hole there is a certain gap, according to experience is usually taken 0.1–0.2 mm is appropriate, so as to avoid the occurrence of interference fit when splicing. Also consider the gap left for the bonding layer, the length of the tab should also be slightly greater than the depth of the positioning hole, generally 1–2 mm. positioning structure is best combined with the printing direction, so that the positioning surface without step effect, forming a better printing plane.

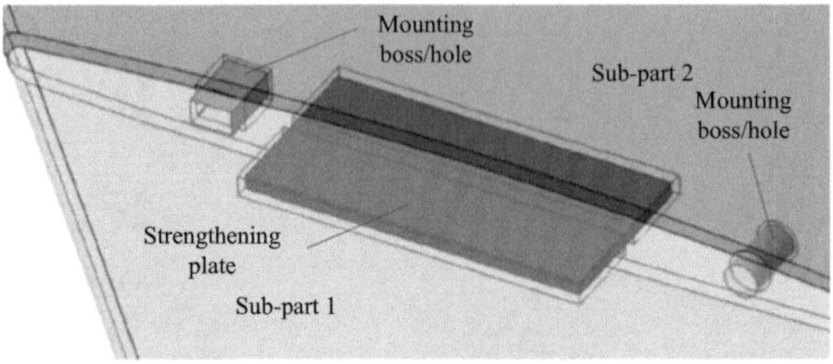

Fig. 2.26 Design of the positioning structure of the split assembly

The partitioning of large wind tunnel test models is a process of continuous optimization around the deformation problem of 3D printing manufacturing. The number of sub-pieces is grasped from the whole, the manufacturing direction of the model is determined by combining with the accuracy requirements, and then a reasonable division position is selected to design the assembly support structure. Before the sub-model is printed and manufactured, support is added according to the pre-considered manufacturing direction. The support should suppress the deformation well, especially to ensure the printing quality at the positioning structure. If the partitioning scheme does not meet the printing requirements, corresponding modifications are made to optimize the manufacturing deformation.

2.3.2 Connection Design of Wind Tunnel Test Model

1) Resin-to-metal connection

The connection of resin parts with embedded metal parts is mainly used to improve the model mechanical properties of the composite reinforcement program, which requires simplified processing (especially metal parts), accurate positioning and firmness. The assembly of resin parts with metal skeleton is mainly done in two ways:

(1) **Resin parts are bonded directly to the metal skeleton surface using an adhesive**

This kind of assembly is mainly applied to resin parts that do not need to be disassembled, as long as the binder is evenly applied to the surface of the metal parts, and then the metal parts are inserted into the resin parts, and after the binder solidifies, the assembly is connected. Since the resin parts cannot be disassembled, the resin parts must be installed successfully at one time, so the processing accuracy of the model is required to be high.

(2) Resin parts are fixed to the surface of the metal skeleton with fastening screws

The resin parts of this assembly method can be disassembled repeatedly. Due to the disadvantages of brittleness and easy wear of resin material, the screws may extrude and wear the resin parts, affecting the quality of the connection surface and the strength of the connection. To overcome these drawbacks and to be able to ensure the reuse of the parts, two options were studied:

The assembly part is embedded with metal parts. As shown in Fig. 2.27 the metal insert is inserted into the aperture and fixed in the resin part by the adhesive. When manufacturing the metal insert, the mesh can be rolled on the outer surface of the metal insert to increase the surface action of the binder and the metal insert to enhance the bonding performance. Since the metal insert is mainly used to bear the fastening force of the screw, the coaxiality of the hole during assembly needs to be ensured in order to avoid serious tilt or misalignment during the threaded connection.

Add metal gaskets to the assembly parts. As shown in Fig. 2.28 the shim is placed in a recess in the resin assembly area to carry the screw fastening force. This solution requires the resin step to have a certain thickness to withstand the large load, while the depth of the recess ensures that the shims and screws are not exposed.

The advantages and disadvantages of the two solutions are shown in Table 2.6. The advantages of the metal insert are that it is suitable for any assembly part, but the assembly is complicated, and the bonding hole has to be polished to remove surface impurities before assembly; the coaxiality of the hole has to be ensured during assembly to prevent the bonding agent from overflowing, unevenness and other reasons from affecting the shape characteristics of the aircraft; since there will be a certain gap between the diameter of the 3D printed hole and the design diameter,

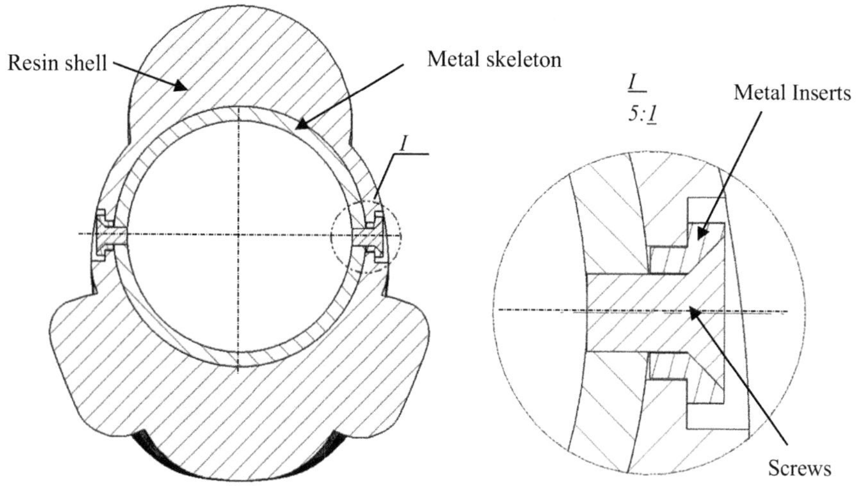

Fig. 2.27 Metal insert solution

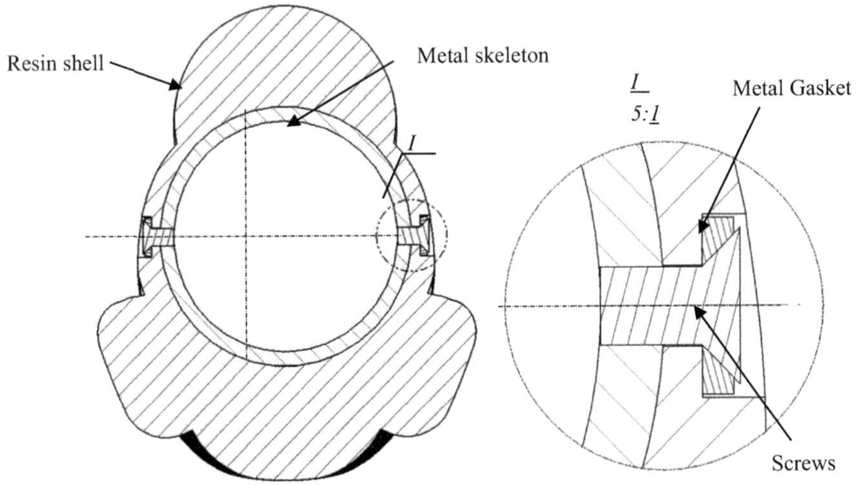

Fig. 2.28 Metal gasket solution

there is a defect that the metal sleeve cannot be properly installed and positioned, and if the actual resin hole diameter is larger than the metal, the radial offset error will be large; if the actual resin hole diameter is smaller than the metal, the resin hole needs to be reamed and reamed, which may also have an impact on the positioning error. The advantage of the shim is that it is easy and fast to assemble and does not require high printing accuracy for the bore diameter and concave table, but the solution has certain requirements for the thickness of the resin step. If the thickness of the step is too small, it will not be able to withstand the large load, and the step will be easily broken or worn through, which will not be able to play the role of screw fastening; at the same time, the depth of the recess should not be too shallow, so that the shim and the screw will not be exposed and destroy the shape characteristics of the model.

Table 2.6 Advantages and disadvantages of screw fastening solutions

Screw fastening solutions	Advantages	Disadvantages
Metal inserts	Suitable for any assembly part	High processing requirements Cumbersome assembly and slow speed
Gasket	Easy and fast assembly	Certain thickness requirements for assembly parts

In summary, considering the simplification of the model assembly link, we choose the option of adding shims as long as the resin thickness of the assembly part can be guaranteed.

2) Resin-Resin connection

The connection between resin parts mainly includes the split to meet the test requirements and the split to meet the printer processing range requirements and improve the printing accuracy of key parts, which requires the corresponding seams to fit well and not affect the pneumatic shape. The connection between resin parts can be divided into the following two options:

(1) **Adhesive connection**

This solution is mainly used for resin parts with non-structural loads and light loads, as well as resin parts that do not require disassembly. The binder is used in the resin parts that need to be connected, and at the same time, structures such as tabs can be designed between the assembly surfaces to increase the contact surface between the binder and the resin parts to improve the bonding performance. Due to the processing error of 3D printing, there may be a certain gap between resin assembly surfaces, which can be repaired by adding resin secondary light curing, as shown in Fig. 2.29.

(2) **Metal reinforced connection**

This solution is mainly used for resin parts with larger loads and resin parts that cannot be easily mounted to a metal skeleton. Under large loads, resin parts connected by binder alone are prone to fracture, so it is possible to consider adding metal-reinforced

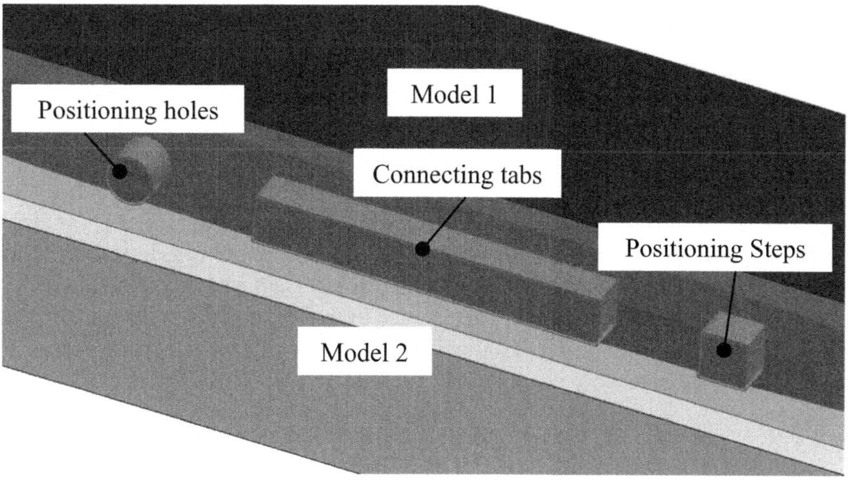

Fig. 2.29 Adhesive connection

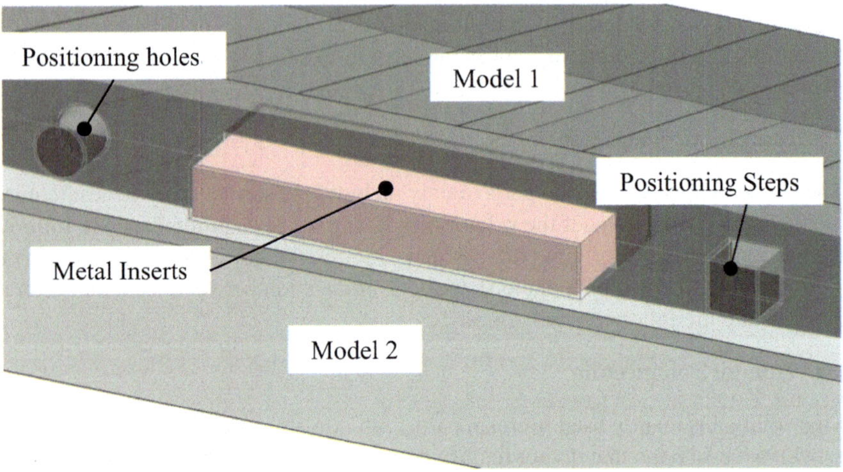

Fig. 2.30 Metal insert connection

inserts between resin parts and connecting the metal-reinforced inserts to resin parts by binder to improve the strength of the connection between resin parts, as shown in Fig. 2.30.

Both resin-metal connection and resin-resin connection need to consider the assembly positioning of the model to ensure the correct installation position of resin parts and metal inserts. Resin material is difficult to adopt pin positioning because it is easy to wear, so positioning structures such as tabs, concave holes and steps can be designed. The number of positioning structures should not be too many so as not to increase the processing difficulty and assembly difficulty, and generally 2–4 are appropriate according to the size of the assembly surface.

When resin parts are assembled with metal inserts, the processing accuracy of both is required to be consistent and the assembly accuracy is good. The key is that through a tight connection, the main load of the model can be transferred to the metal parts to ensure the safety of the resin parts, so ensuring the tightness of the two connections and the coordination of deformation is the key to such connections. Due to the resin material characteristics and the processing characteristics of light-curing 3D printing, resin parts are easy to shrink when printing holes, slots and other structural shapes, and there is a certain loss of dimensional accuracy of the parts, which is easy to produce an interference fit in the assembly, making it difficult or impossible to assemble resin parts. Therefore, the following two measures are considered in the assembly design:

(1) to leave a reasonable assembly gap between the assembly structure, according to general processing experience, resin parts and metal inserts assembly, assembly gap is usually 0.2–0.3 mm appropriate; if resin parts and metal inserts need to use binder, assembly gap is usually 0.4–0.5 mm appropriate;

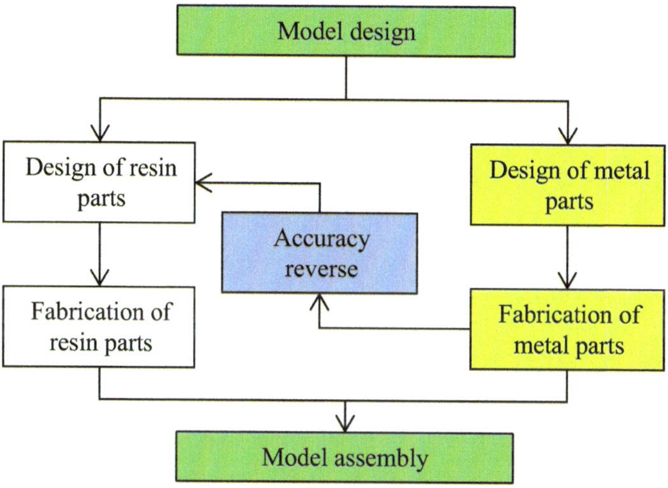

Fig. 2.31 Technical solutions to ensure the accuracy of resin-metal connections

(2) The machining steps shown in Fig. 2.31, the metal parts are manufactured first in the model design, and the dimensions of the resin parts are modified according to the inverse results to optimize the assembly effect.

Chapter 3
Process Basis of Additive Manufacturing for Wind Tunnel Test Models

3.1 Relevance of Design and Manufacturing for Wind Tunnel Test Models

In terms of processing range, the existing 3D printing equipment has the limitation of processing size and cannot manufacture the whole wind tunnel test model with larger size or large size wind tunnel test model parts, which need to be designed for splitting the wind tunnel test model. In terms of accuracy characteristics, 3D printing manufacturing is not limited by the complexity of the shape and structure of the parts, which has strong advantages for manufacturing wind tunnel test models with complex streamline shapes and complex internal structures. However, because the processing accuracy of 3D printing is $\pm$ 0.1 mm, to a certain extent, it will cause the absence of tiny features, and there are limitations for the microfine structural features of wind tunnel test models such as ultra-thin wingtips and ultra-small deep apertures, thus the design of wind tunnel test models can be modified to a structure suitable for 3D printing processing under the premise of meeting the aerodynamic shape and testing requirements. In addition, because the step effect will have a large impact on the aerodynamic shape accuracy of the wind tunnel test model, the wind tunnel test model shape should be designed to compensate for the different printing directions. In terms of manufacturing economy, the manufacturing cycle and manufacturing cost of 3D printed parts are not related to the shape and complexity of the parts, but only to their net volume, and the manufacturing cost and processing time can be estimated based on the volume of the CAD model. In addition, to ensure the strength and stiffness of the model, other low-cost and high-performance materials can be considered for filling inside the wind tunnel test model.

W. Zhu and D. Li, *Models for Wind Tunnel Tests Based on Additive Manufacturing Technology*, Additive Manufacturing Technology,
https://doi.org/10.1007/978-981-99-5877-1_3

3.2 Additive Manufacturing Process for Wind Tunnel Test Models

The resin parts of the model were processed on the SPS600B light-curing rapid prototyping machine at Xi'an Jiaotong University. SOMOS 14120 was selected as the resin material, and Magics RP 10.0 software was used to prepare the data before light-curing printing, including the selection of the printing direction of the resin parts, the design of the added support, and the layout design.

3.2.1 Data Pre-processing

1) Data Conversion

At present, almost all types of 3D printing systems use STL data format, which is the data exchange standard between CAD system and 3D printing system, it discrete the continuous surface of 3D CAD model into the geometry of triangular pieces, so it greatly simplifies the data format of 3D CAD model, but STL data format only approximates the outer surface of 3D CAD model, the smaller the size of small triangular pieces, the more the number, the higher the accuracy of the model approximation. The smaller the size and the more the number of small triangular slices, the higher the accuracy of the approximate approximation of the model. The smaller the size and the more the number of small triangles, the higher the accuracy of the model approximation. Generally, in the 3D CAD software, when exporting STL format files, the accuracy parameters are required to be entered, which is the maximum allowable error of fitting the original CAD model with STL format. However, for surface models, no matter how to improve the conversion accuracy, the small triangular face piece can not completely fit the surface, especially the larger the radius of curvature, the more obvious the error. For a model such as shown in Fig. 3.1, the exported data will be closer to the prism if the accuracy parameters are set to a limited extent.

The fundamental way to eliminate such errors is to use the CAD data directly in the 3D printing system without STL conversion. But so far there is no technology that can reach this step, the existing way is to rely on experience to set the appropriate accuracy parameters to reduce the error when STL format conversion of CAD models. Since most of the aircraft wind tunnel test models are curved surfaces, the surface accuracy of the shape is very critical, so the data design from CAD to STL has an important impact, and the common way of outputting the file directly to STL format is not desirable, which largely reduces the accuracy requirements of the printed parts.

The model is designed in the design software, and the triangle tolerance can be set when exporting the STL file format. Figure 3.2 shows the effect of STL triangular surface delineation under different accuracy conditions. From the figure, it can be seen that if the triangle tolerance is set larger, the surface of STL data will have obvious stripes, which will affect the surface accuracy of the model and thus reduce the printing accuracy; if the triangle tolerance is set smaller, the surface of STL data

Fig. 3.1 STL file format for cylinders

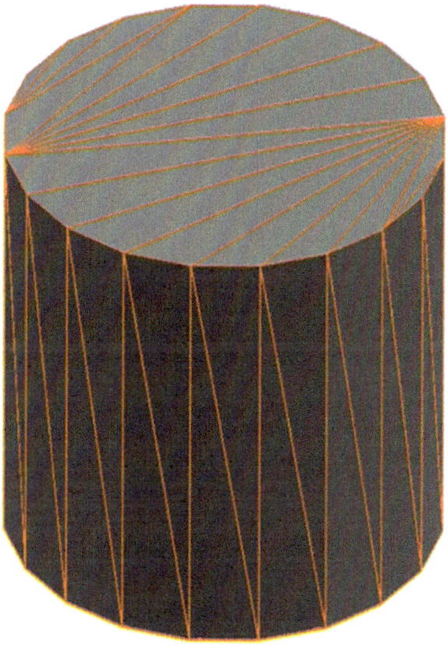

will be smoother, which ensures the printing accuracy of the model. But in the CAD output STL, not the higher the accuracy is better, this is because the characteristics of the 3D printing process itself determines its accuracy in ± 0.1 mm, too high precision requirements exceed the accuracy indicators that can be achieved by the 3D printing manufacturing system, generally huge amount of triangle face piece will reduce the running speed of Magics RP 10.0 software, the processing time of the model face piece increases significantly, at the same time huge amount of the large number of triangular facets will create many small line segments in the model profile cross-section, which is not conducive to the scanning motion of the beam, resulting in low productivity and poor surface finish.

The accuracy of the aerodynamic profile of the wind tunnel test model affects the accuracy of the wind tunnel test data, especially the printing accuracy of the nose tip, leading edge of the front flap, trailing flap, trailing edge of the aileron and other curved parts is especially important. This not only ensures the accuracy of the data but also minimizes the size of the data file and improves the efficiency of subsequent data processing.

2) Offset Compensation

Since 3D printing technology is an ensemble of 3D entities discrete into cut layers with thickness, model layering not only destroys the continuity of the model surface, but also loses the contour information between adjacent cut layers. The thickness of the layers is the resolution of the discrete model, the larger the thickness of the layers, the lower the resolution, the more data the model loses, and the greater the printing error. The following two main types of errors are likely to occur during model printing:

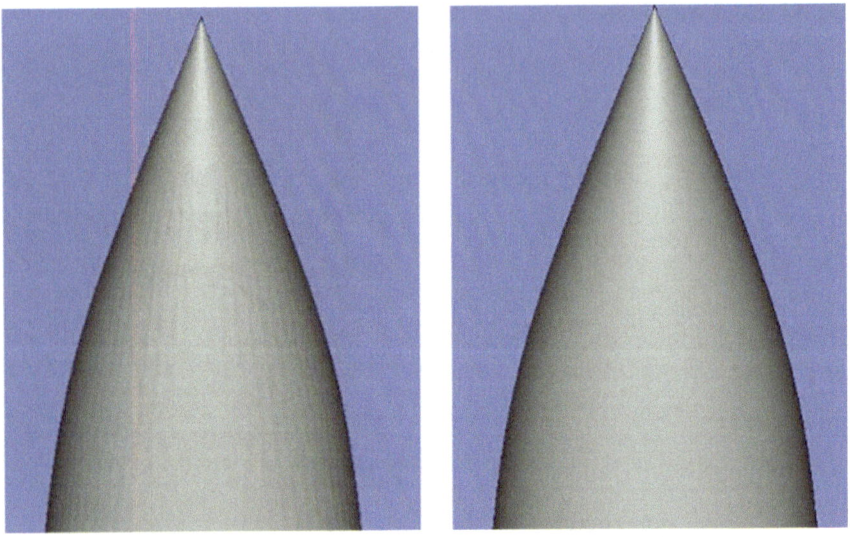

(a) Data conversion with an accuracy of 1.0mm (b) Data conversion with an accuracy of 0.08mm

Fig. 3.2 STL triangular surface slice division under different accuracy conditions

Layering error: When the model is layered, each layer always has a certain thickness, set as Δt, the model along the cut layer direction of the dimensional length of l, when l can not be divided by Δt the model will have a part of the thin layer can not be printed, it will cause the dimensional error in the direction of layering. For example, a cone with a tip, such as Fig. 3.3. If the last layer slice is located inside the solid, i.e., the thin layer of the model is lost, which results in the model decreasing the dimensional accuracy of Δl in the tangent direction; if the last layer slice is located outside the solid, i.e., the thin layer of the model is exaggeratedly expressed, which results in the model increasing the dimensional accuracy of $\Delta t - \Delta l$ in the tangent direction increases the dimensional accuracy of $\Delta t - \Delta l$.

Step error: When the model is layered, the shape profile of the model consists of two tangent layers, and since the horizontal surface profiles of the upper and lower tangent layers are not the same, the surface profile of the model must be replaced by the column surface profile, as in Fig. 3.4. The surface accuracy error caused by 3D printing process is related to the normal direction of the solid surface, radius of curvature and the thickness of the layered surface, which can be improved by reducing the thickness of the layered surface and optimizing the manufacturing direction. In 3D printing, the surface accuracy of the part can be improved by reducing the layer thickness and optimizing the manufacturing direction. However, the shape of the aircraft model is complex, especially the wing part of the aircraft, and the accuracy requirements of different parts are different. In addition, the resin printing process mainly relies on the polymerization chemical reaction of monomer molecules, the resin inevitably produces volume shrinkage [35], especially with the metal skeleton

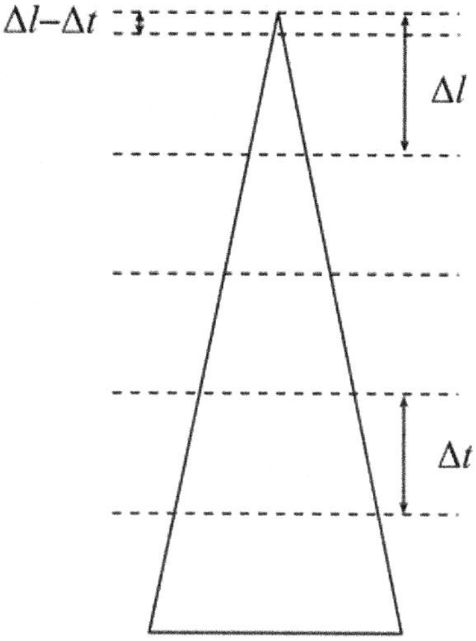

Fig. 3.3 Stratification error

with holes, slots and other structures, shrinkage size is large, so it is also necessary to offset compensation design for such structures. The offset compensation should be combined with the actual size of the skeleton to make fine adjustments as appropriate.

For example, the tip of the head, its thin tip part is likely to cause the loss of processing features due to the influence of 3D printing manufacturing process, forming a certain curvature of rounded corners in the printing process, which not only changes the shape characteristics of the head, but also reduces the length size of the

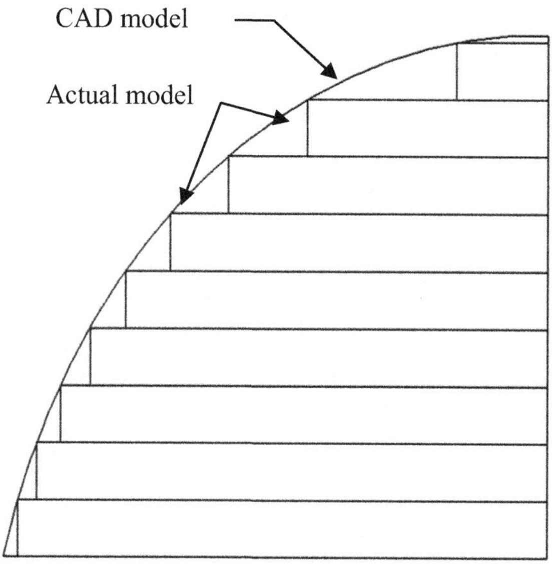

Fig. 3.4 Step error

head along the x-axis, so the shape characteristics of the tip part of the head are shifted outward in the design, which can not only compensate for the processing errors of 3D printing, but also offset the model. The amount of post-processing resharpening, as shown in Fig. 3.5. As shown in Fig. 3.5, the front end of the head is selected to be 5 mm and offset by 0.2 mm.

For example, the front flap of the wing, because its wing section is printed at an angle of 50°, the step effect causes the leading edge of the flap to have a certain dimensional contraction in the horizontal plane direction, in order to ensure the dimensional accuracy of the leading edge we offset the leading edge outward by 0.2 mm, as shown in Fig. 3.6.

For example, the nose, front fuselage, left and right wings and tail parts, the inner diameter value and the metal skeleton outer diameter value generally have a gap of 0.4 mm. Since the metal skeleton outer diameter value is slightly reduced, the offset

Fig. 3.5 Head compensation

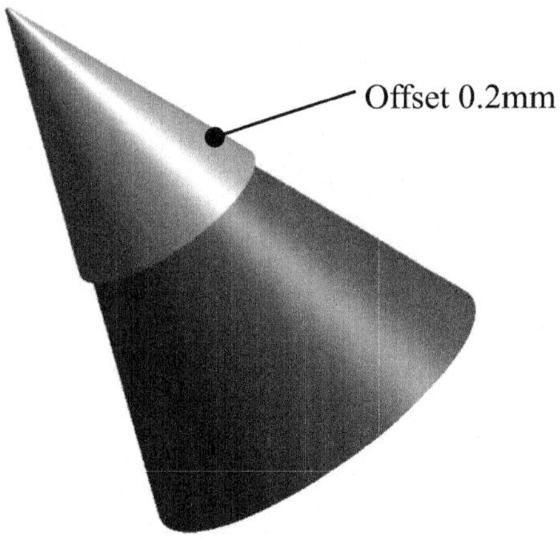

Fig. 3.6 Front flap compensation

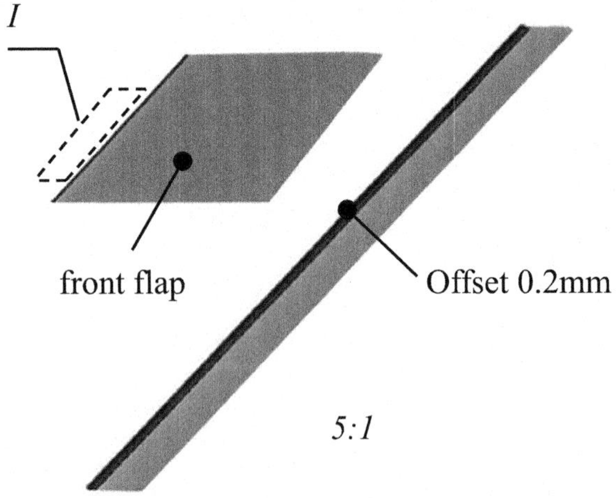

$\delta_1 = 0.3$ mm is taken in the resin inner diameter compensation design, as shown in Fig. 3.7. In addition, the deviation at different plate thicknesses after processing of the wing plate is also different, so the deviation of the resin should be adjusted accordingly, as shown in Fig. 3.7. As shown in Fig. 3.8, the offset of the flat plate is generally taken as $\delta_2 = 0.25$ mm, the offset of the upper surface of the flat plate at 6 mm plate thickness is taken as $\delta_3 = 0.35$ mm, and the offset of 4 mm plate thickness is taken as $\delta_4 = 0.3$ mm.

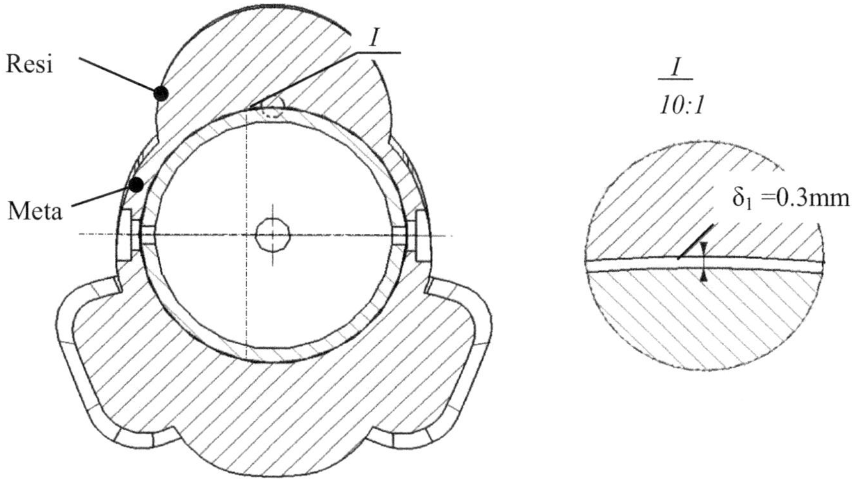

Fig. 3.7 Resin head—metal head fit compensation

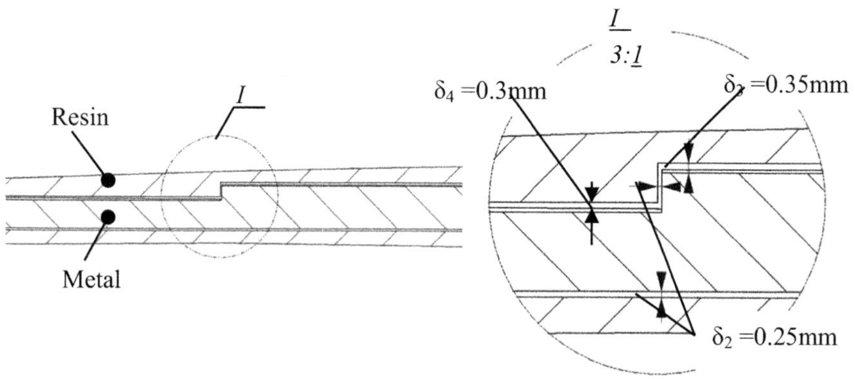

Fig. 3.8 Resin wing—metal wing fit compensation

3.2.2 Print Process Selection

1) Print direction determination

In 3D printing manufacturing, the printing direction of the part has a greater impact on the printing accuracy of the part [36], especially the aircraft model requires high printing accuracy on its surface. In order to minimize the impact of the step effect on the manufacturing accuracy of the model surface, different resin parts should choose different printing directions.

As shown in Fig. 3.9, the printing direction of the resin nose, front fuselage, wings and tail is chosen in the direction of the longitudinal axis of the model, which can ensure the manufacturing accuracy of the surface of the model shape and make the flow of the fuselage shape consistent, and also ensure the manufacturing accuracy of the inner axis holes of the model, and ensure the assembly accuracy of the resin nose, front fuselage, wings and tail with the metal skeleton.

As shown in Fig. 3.10, the print direction of the model's maneuvering surfaces have to be set carefully. In the force measurement model, the leading and trailing edges of the maneuvering surfaces are the key parts that have a large impact on the aerodynamic performance. In order to maximize the shape accuracy of the edges, the printing direction of each maneuvering surface is generally set to the direction of the wingspan. For example, the front flap, aileron and rear flap in the main wing are printed along the wingspan direction of the main wing, while the flat tail model and rudder model are printed along their wingspan direction respectively.

2) Support structure design

In the light-curing printing process, the support structure can be considered as a tooling central tool that is manufactured at the same time as the prototype part in order to ensure the precise positioning of the prototype part relative to the machining

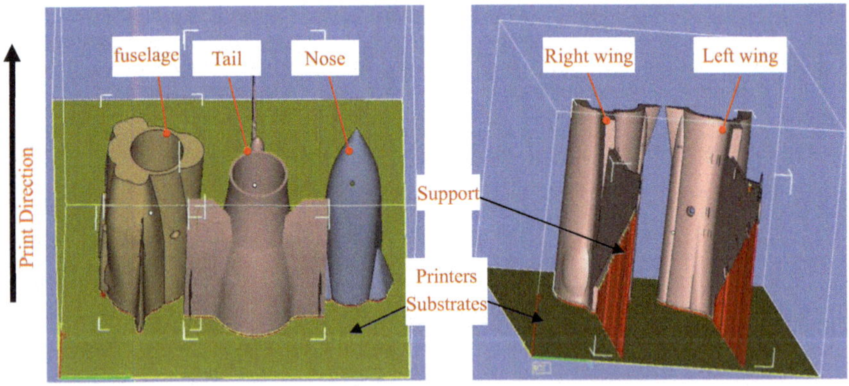

(a) Fuselage printing (b) Left and right wing printing

Fig. 3.9 Schematic diagram of resin shell processing

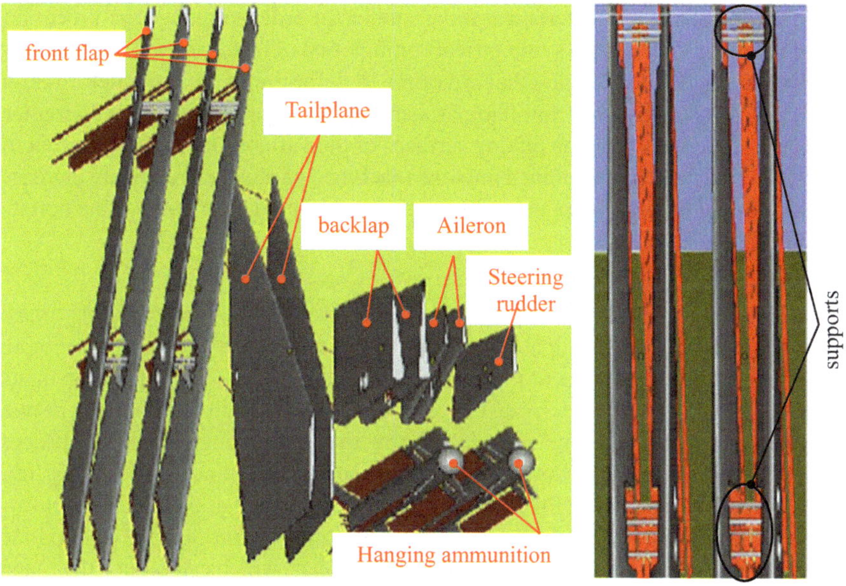

(a) Maneuvering surface and hang-up printing (b) Front flap printing

Fig. 3.10 Schematic diagram of resin manipulation surface and hang-up processing

system during manufacturing, while isolated contours and cantilevered contours in the part also need to be positioned by the support structure. In addition, supports must be added to the bottom of the printed part in order to make it easy to separate the print from the table without damaging the part. Supports can be divided into cross supports (is one of the most common support methods for general features and areas of internal filling), polygonal supports (with better stability and strength), oblique supports (cantilever supports for cantilever structures), hand-painted support whisk (generate supports as needed). The design principles of supports are as follows:

(1) To ensure the printing accuracy and print stability under the premise of minimizing the support added area;
(2) Adding auxiliary support to balance shrinkage stress and reduce part deformation;
(3) The support should be easy to remove and reduce the breakage of the part surface.

Figure 3.10a shows, the support of the model front flap, aileron, rear flap, flat tail and rudder are chosen on their assembly surfaces, and the assembly surfaces are all flat, which ensures the shape accuracy of the edges of each maneuvering surface and facilitates the post-processing repair of the assembly surfaces. For the long and thin parts such as the front flap, when processing and manufacturing in the printing direction along the vertical wing span, the front flap is easy to produce the arch deformation along the wing span due to the resin shrinkage stress, thus changing the leading edge shape characteristics of the model. In order to reduce the processing deformation of such parts, a symmetrical combination of layout is used, and auxiliary supports are added to the parts that are prone to deformation or larger deformation. As Fig. 3.10b shows, two front flap parts are laid out symmetrically, and auxiliary support is added between the mating surfaces of the variable angle piece, so that the two machined parts are combined into one machined part, which relatively increases the stiffness of the front flap and reduces the machining deformation of the parts.

3) Layout design

As mentioned above, 3D printing can significantly reduce the manufacturing time of wind tunnel test models compared to traditional machining manufacturing methods. Nevertheless, due to the size of the part and the efficiency of 3D printing, the manufacturing time for most models can exceed a dozen hours, with tens of hours or even days for larger parts (e.g., meter-scale). Taking the light-curing 3D printing process as an example, the manufacturing time is mainly reflected in the laser scanning time, squeegee movement, pallet lift and liquid level stabilization assisted manufacturing time. In the auxiliary time, there is no actual part manufacturing, and the auxiliary time of one part and multiple parts is basically similar, which can reduce the manufacturing time by multiple parts manufacturing [37]. Aircraft models require high form accuracy, so the printing direction of model parts is often chosen to cut the direction with the largest number of layers, and the increase in the number of layers inevitably increases the overall manufacturing time.

The manufacturing of the aircraft model should be made with the highest goal of manufacturing accuracy and the secondary goal of reducing the manufacturing time of the model, and the grouping and layout of the model should be developed. The aircraft model can be divided into fuselage shape part, wing shape part and maneuvering surface part according to the similarity of the model. Within the processing range allowed by the printer, the similar model parts are best printed together, using the same scanning speed, layering thickness and other process parameters to form similar processing errors. The resin parts of this model were manufactured in three stages:

(1) The fuselage shape part: nose, front fuselage, tail;
(2) Wing profile part: left and right wings;
(3) Maneuvering surface parts: left and right 0° front flaps, left and right 24° front flaps, left and right rear flaps, left and right ailerons, left and right flat tails, left and right hang-ups and rudders.

3.2.3 Manufacturing and Post-processing

The model is manufactured using the SPS600B light-curing printer developed independently by Xi'an Jiaotong University. The printing range is 600 mm × 600 mm × 450 mm, and the printing accuracy is ± 0.1 mm. The main process parameters are described in Chap. 1.

Fig. 3.11 Resin wing processed in the printer

Figure 3.11 shows the finished resin wing processed in the printer. After the resin part is manufactured, it needs to be raised and removed from the liquid photosensitive resin tank, in which the surface will adsorb liquid resin, which will not only destroy the surface finish of the part after curing, but also may lead to dimensional and shape errors of the part. Therefore, when the part is manufactured and the pallet is raised, it should wait for a period of time before removing it, so that the liquid resin adsorbed on the surface of the part can flow back into the resin tank and also save Resin. Parts must be cleaned in time after removal to avoid curing of the adsorbed resin on the part surface. Use other tools such as a razor blade to strip the support from the part while cleaning it with alcohol, and then sand it afterwards.

3.2.4 Model Testing

For 3D printed and assembled resin parts, the structural integrity and aerodynamic profile smoothness are first checked by observation after gap filling (e.g. putty) and surface treatment (e.g. sandpaper sanding). For more demanding models, the dimensional accuracy and surface roughness should also be checked.

According to the requirements of GJB 180A-2006 A "Low-speed wind tunnel aircraft model design specification" [4], the accuracy of the fuselage is required as follows

(1) The error between the head surface of the fuselage and the reference is not more than 0.08 mm;
(2) The error of other parts of the body and the benchmark is not more than 0.1 mm;
(3) Length: ± 1.0 mm.

3.3 3D Printing Error Analysis and Compensation

Wind tunnel test models especially need to ensure the aerodynamic shape, so there are high requirements for surface errors, and the step effect and model post-processing when using 3D printing manufacturing will have a significant impact on the surface quality, but certain corrections can be made in the design to minimize the corresponding errors. Therefore, the design for 3D printed manufacturing can make structural corrections and design compensation for the surface errors caused by the step effect and the foreseeable post-processing errors.

3.3.1 3D Printing Related Error Analysis

1) Layered processing error analysis

Since 3D printing is a manufacturing method based on the principle of material accumulation, that is, the 3D entity is discrete in the direction of material accumulation into a finite number of cut layers with thickness. After the layering process, the original continuous surface of the model in the layering direction is discrete. As a result of layering, only the contour information and solid information of each slice layer are obtained, while the outer contour surface information of two adjacent slice layers is lost, which is an approximate representation method. The thickness of the layers indicates the resolution of the model expressed after discretization, and the larger the thickness of the layers, the lower the resolution, the more information is lost, and the greater the error generated by the printing process. In particular, the surface inclined relative to the printing direction and the surface shape, due to the presence of the step effect, makes the surface accuracy significantly reduced, in addition, the step effect is also a factor that causes post-processing errors.

Currently in 3D printing technology, the indicators for evaluating the degree of step effect are generally ε and δ [24]. As shown in Fig. 3.12 ε is the maximum distance between the layer-piece stacked boundary and the original boundary of the CAD model; δ is the maximum distance between the original boundary of the CAD model and the boundary of the layer-piece stacked solid measured along the direction of the surface normal vector of the CAD model. ε value can better describe

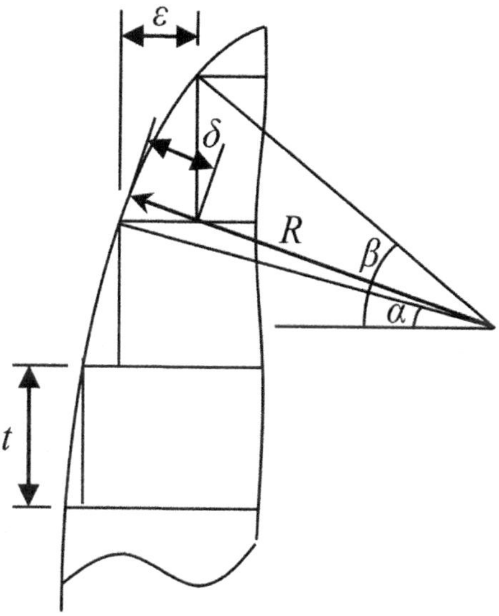

Fig. 3.12 Schematic diagram of step effect error

the volume error of the solid, while δ value can better correspond to the surface roughness of the solid. Let the total number of solid layers be n, then max $\sum_{i=1}^{n} \varepsilon_i$ and max $\sum_{i=1}^{n} \delta_i$ characterizes the maximum volume error and surface roughness of the part, respectively.

Let the radius of curvature of the wind tunnel test model surface at a point on the surface of the current layered height entity be R. For convenience, the radius of the contour arc between the current layered thicknesses can be approximated as equal to R. Let the angle between the normal vector at the beginning of the arc and the stacking plane be α, the angle between the normal vector at the end and the stacking plane be β, and the layered thickness be t, as in Fig. 3.12 shown, then:

$$\sin\beta = \sin\alpha + t/R \tag{3.1}$$

$$\varepsilon = R\left[\cos\alpha - \sqrt{1 - (\sin\alpha + t/R)^2}\right] \tag{3.2}$$

$$\delta = R - \sqrt{t^2 + R^2 - 2tR\sin\beta} \tag{3.3}$$

When $\alpha = 0°$, $\varepsilon = \delta = R\left[1 - \sqrt{1 - (t/R)^2}\right]$, i.e., the two error evaluation metrics are identical.

When $\beta = 90°$, $\varepsilon = R\cos\alpha$, $\delta = R - \sqrt{(t - R)^2} = t$. That is, the maximum distance between the CAD model surface and the solid boundary along the direction normal to the solid surface at this time is the delamination thickness.

It can be seen that the profile accuracy errors generated by the 3D printing process are related to the normal direction of the solid surface, the radius of curvature, and the layered thickness.

1) Post-processing error

After the printed workpiece is removed from the printer, it needs to be stripped of its support structure, and in some cases, post-curing, repairing, grinding, polishing and surface treatment, which are collectively referred to as post-processing. In this process, if not handled properly will affect the size and shape accuracy of the prototype, resulting in post-processing errors. Post-processing can be divided into the following two types:

(1) Post-processing to improve surface quality: After the workpiece is printed, not only does it need to be removed from the support, but it also needs to be repaired, sanded and polished, for example, the surface of the part is not smooth, and there are small steps and defects on the surface caused by layered manufacturing.
(2) Post-treatment to improve surface properties: The surface condition and mechanical strength of the parts are not fully satisfied with the requirements of the final product, and the workpiece may continue to deform and cause errors due to changes in temperature, humidity and other environmental conditions or the influence of printing residual stress. The surface color of the product can be changed or its strength and other properties can be improved by surface coating. The electrodeposition manufacturing described later in this book can also be considered as a post-treatment process.

3.3.2 Offset Compensation Design for 3D Printing

Different placement angles when 3D printing will result in different types of surfaces at different locations on the model. Upward horizontal surfaces do not produce a step effect and do not require support; downward horizontal surfaces do not produce a step effect, but the larger the total area of such a surface, the greater the likelihood that support will be required or the more support will be needed; upward sloping surfaces produce a step effect and do not require support; downward sloping surfaces produce a step effect and may or may not require support, depending on the overall shape of the surface Whether support is needed depends on the overall shape of the surrounding; vertical surface does not produce a step effect, and generally does not require support; near-vertical surface produces a small step effect, and generally does not require support.

As different surfaces have different errors, the step effect formed by the upward sloping surface makes the actual model present a decrease in material compared with the theoretical model, i.e., the error is negative deviation; while the downward sloping surface presents an increase in cured material, i.e., the error is positive deviation. The shape of aircraft wind tunnel test models are curved design, especially the wing has a great impact on the aerodynamic performance, so the shape of the wing processing to ensure a high degree of accuracy. Generally 3D printing manufacturing single parts in order to save printing time, the parts and the horizontal plane into a conventional 10°–30° printing, such as Fig. 3.13. Wing design generally has a chord plane will be divided into the shape of the upper and lower airfoil, the upper and lower surfaces show different error direction, making the overall wing surface deviation, and step effect error is larger, will increase the post-processing error, should be upward sloping surface for some design compensation. As shown in Fig. 3.14, if the wing is printed at an angle of 30° from the horizontal, the angle between the normal and vertical directions of the surface is slightly less than 30°, and the error δ is about $0.8t$. The model obtained by processing the upper surface after offsetting it by $0.8t$ will be close to the theoretical boundary after removing the step effect.

When 3D printing, the use of different machining directions creates different form positions and thus different accuracy requirements. Because different models and different parts on the same model also have different accuracy requirements. Generally, high-speed aircraft have higher requirements than low-speed aircraft; wing parts have higher requirements than fuselage-type parts on the same aircraft; the part before the maximum section has higher requirements than the part after the maximum section on the same part. The printing direction of the model can be divided or arranged according to the accuracy requirements of different parts, and the wind tunnel test model can be designed with certain compensation by combining different printing directions. If the processing is done along the wing span direction in 3D printing, so that the surface of the wing forms an upward approximated vertical surface when 3D printing, as Fig. 3.15 shown, the surface accuracy of the model can

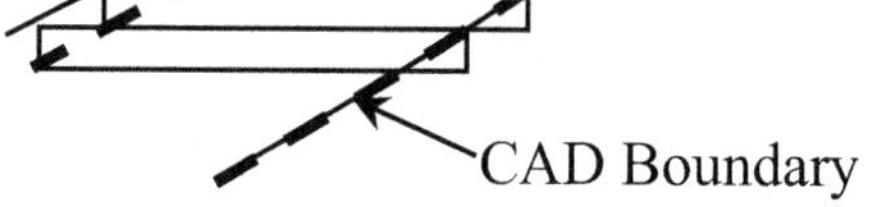

Fig. 3.13 Effect of step effect on the shape surface

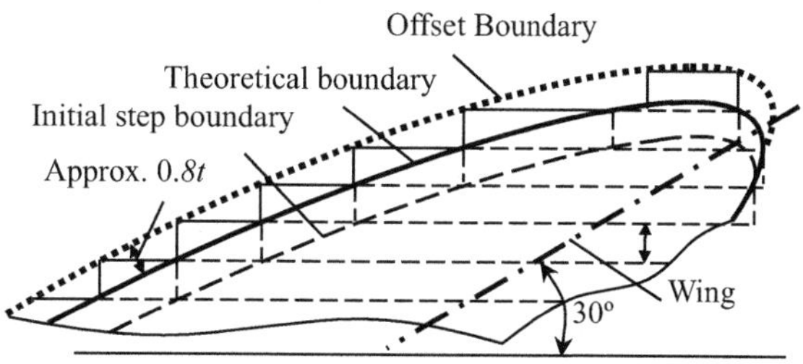

Fig. 3.14 Offset compensation towards the upper surface

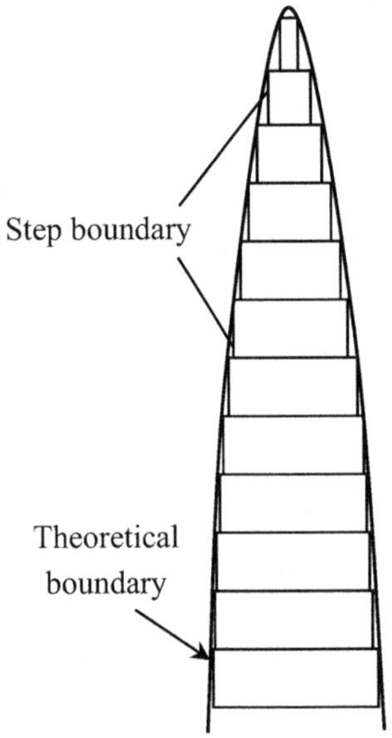

Fig. 3.15 Step effect of vertical manufacturing

be improved. In addition, the printing errors of each key surface are in the same direction with negative deviation, which facilitates the overall offset compensation. However, in the wing tip due to the transition from horizontal to vertical curvature direction, a larger step effect will be generated, and the local enlargement of this part can provide a corresponding ε offset to the cut layer profile, forming a better form of error compensation, which is calculated schematically as shown in Fig. 3.16. If the offset compensation has been designed for the overall model so that the model tends to have positive deviation in general, the amount of grinding in the curvature transition area can be properly controlled in the post-processing process to form a better surface quality and profile accuracy.

The thin-tip feature or even zero-thickness area is commonly found in the wings of aircraft wind tunnel test models, and the thin-tip part of its tail is likely to be lost due to the influence of 3D printing manufacturing process, or the wing tip forms a jagged shape instead of a straight line, so in order to maintain the thin-tip feature, the offset design cannot only compensate the error caused by the step effect, but also should increase some compensation amount for the linear feature formed by the trimmed wing tip. Therefore, it is necessary to cut the wing tip of the revised model to a position that tends to the theoretical line to ensure the position control when resharpening, as shown in Fig. 3.17.

In the design of the model, the maximum volume error of the upward sloping surface can be compensated by local offset. As the post-processing removal of the support and step effect still needs to smooth the grinding amount of the model surface, it is necessary to set aside about 0.1 mm for grinding. The metal wind tunnel test model is generally reserved with 0.1 mm margin. The resin model is easier to operate than the metal one, so a larger amount of grinding can be reserved to give a freer

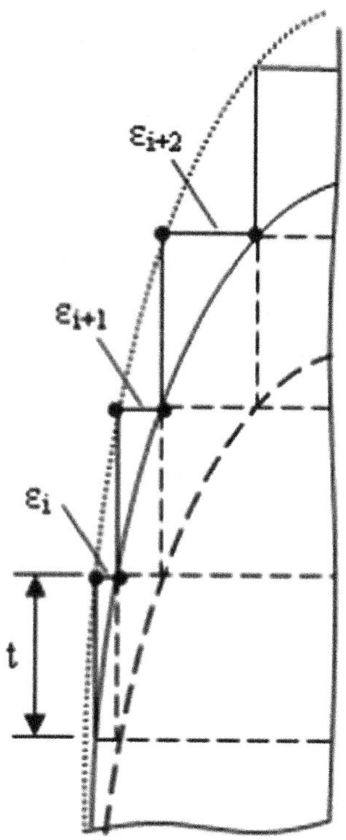

Fig. 3.16 Offset compensation schematic

margin for post-processing, so for 3D printing, the surface offset compensation value is about 0.3–0.5 mm. The design of the model compensation can be considered during the design of the wind tunnel test model, as well as the design of the CAD surface or the appropriate editing of the converted STL wind tunnel test model.

1) CAD Offset Design

Different surfaces have different spatial structures and different connection methods (point continuity, slope continuity, curvature continuity) between surfaces, while different thicknesses and different directions also produce different offset effects, and different methods can be used for offset design of CAD surfaces. The offset compensation direction for 3D printing manufacturing faces outward along the curvature direction to compensate for the missing model caused by the step utility, and the offset thickness is very little, about 0.3–0.5 mm, so it is feasible to modify the CAD model directly. It is possible to first merge (Join) multiple surface pieces into a composite surface, where the surfaces must be connected, and then thicken them together (Thick Surface). In case of failure to thicken the surface due to gaps, sharp corners, or too small transition rounding, the quality of the surface can be checked and

Fig. 3.17 Thin wingtip feature retention design

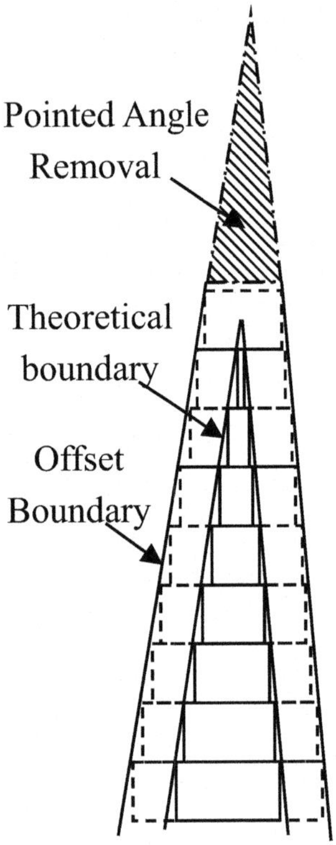

the surface can be repaired using the HA (Healing Assistant) module before thickening, or the group of surfaces with good continuity can be extrapolated, trimmed, and sewn together. The surface composition of the wind tunnel test model is usually the aerodynamic profile and assembly combination plane, which can be compensated separately by offset and then synthesized into a solid by Boolean operation.

2) STL Offset Design

If the CAD model is not corrected, the STL model can be manipulated for offset compensation design. The STL 3D printing (STL Rapid Prototyping) module in CATIA V5 software provides advanced tools to improve mesh quality by removing or reorganizing triangular cells, filling holes and re-meshing the mesh in whole or in part. STL files can be generated quickly and accurately by partitioning CAD data into meshes; existing STL files can be imported to display the mesh and analyze its quality; trace offset meshes can be generated as solids and the mesh can be partitioned and merged; the generated mesh can be exported as a standard binary STL file to be provided to 3D printers. The generated mesh can be exported to a standard binary STL file to be provided to the 3D printer. Also, common data pre-processing software for 3D printing such as Magics RP can edit the STL for compensation design.

3.3.3 Electrodeposition-Oriented Model Correction Design

1) Necessity of electrodeposition manufacturing of resin models

Under normal circumstances, the wind tunnel test model is transported from the manufacturing unit to the test unit for testing, and the use of the wind tunnel test model should consider the storage and transportation of the model. When designing the model, the corresponding packing box must be designed, and the number and volume of the packing box should be decided according to the specific situation of the disassembled parts of the model. The model parts should be placed in such a way that the deformation of the parts in the storage and transportation process is minimized, and in order to prevent the model from being damaged in the transportation and handling process, the model in the packing box should be protected from moisture and rust. In addition, the front surface of the model may be scarred due to the impact of dust particles in the air after long-term use.

Since the resin models obtained by 3D printing are non-conductive, non-thermally conductive, non-abrasive, easily deformable, non-pollution resistant and lacking metallic luster, thus limiting the scope of use to some extent, electrodeposition is considered to improve the limitations of resin models. Electrodeposition is the process of electrochemically reducing metal compounds to metal on the surface of metallic and non-metallic models and forming a smooth and dense metal layer that meets the requirements of. The desired layer thickness can be obtained by controlling the process conditions (plating time, current density, etc.) during electrodeposition. If a layer of metal is plated on the resin, its serviceability can be greatly improved, increasing functionality on the one hand and protection on the other. The main advantages are as follows:

(1) Improving the hardness, stiffness, strength and other properties of the resin model surface;
(2) To make resin models with wear resistance and electrical conductivity and high temperature oxidation resistance, etc., to extend their service life;
(3) To make the resin model highly stable to external factors such as light and atmosphere, which can prevent aging;
(4) Make the resin model surface with metallic luster, beautiful and not easy to contaminate.

2) Model design of resin electrodeposition

The simpler the shape of the model to be electrodeposited, the better. The geometry of the model will have a great impact on the quality of the plating, so the model should be designed with some attention to the following issues:

(1) To avoid the use of large surface plane. Use a slightly curved shape with a surface designed to be arched with a central arch.

(2) To avoid right angles and sharp corners. Angular parts are prone to local current density increase, plating thickening and easy to cause nodular phenomenon. Therefore, the square profile is changed to curved profile as far as possible, or transition with rounded corners.
(3) Do not have too deep blind holes. These parts are not only difficult to plating, but also easy to contaminate the solution of the next process with residual solution. The bottom of deep holes or non-cylindrical deep cavities should be designed as spherical, with small drainage holes in the middle and upper part to make the plating solution circulate, which is conducive to uniform plating. Narrow and slender holes should be changed into through holes as far as possible, which may not necessarily form a plating layer, but facilitate the cleaning of the aperture after plating.
(4) The thickness should not be too thin, and there should not be sudden changes. Too thin parts are easily deformed by heat or plating stress during the plating process; sudden changes in thickness are likely to cause stress concentration, and generally speaking, the thickness difference should not exceed twice.
(5) Leave the necessary plating process holes. Easy to load and unload conductive fixture, easy to hang on the plating tank cathode, and have a large conductive contact surface. Hanging position is designed in the part that does not affect the appearance, and pay attention to prevent the deformation of thin-walled parts.

The shape of the aircraft is characterized by a streamlined curved shape, and the unfavorable shape for electrodeposition mainly lies in the corners of the assembly parts, the wingtips of each maneuvering surface and the pressure measurement micro-apertures on the pressure measurement model. The corners can be rounded transition, the wingtip parts and the pressure measurement micro-aperture channel should be designed and retained in conjunction with the process to avoid affecting the aerodynamic performance. Since the plating layer has a certain thickness, after the surface treatment of the parts, it will inevitably cause changes in the dimensions of the parts. Usually the model dimensions and tolerances specified on the design drawings refer to the final dimensions and tolerances of the model before plating, and the thickness of the plating and the dimensional deviation of its plating should be reserved in advance in conjunction with the final dimensions. Thus the relationship between the assembly interface is not only with twice the deposition thickness, there is also the influence of the assembly error, and the model offset should be considered as a whole when designing. The general electrodeposition thickness is 0.1–0.2 mm, and the 3D printing oriented shape offset compensation is 0.3–0.5 mm, then the 3D printing based electrodeposition offset compensation is about 0.1–0.4 mm.

Chapter 4
Inspection Techniques of Wind Tunnel Test Models Based on Additive Manufacturing Wind Tunnel Test Models

4.1 Manufacturing Requirements for Wind Tunnel Test Models

4.1.1 Machining Accuracy and Surface Roughness Requirements

Since the model and the real object must be geometrically similar, the shape of the model must be precisely controlled when machining it. Usually, the shape of the model is mostly curved, and the existing method to ensure the machining accuracy is through the prototype. The location and number of samples are determined by the complexity of the model shape. Generally speaking, the curvature of the large changes in the place to take a few more samples, the small changes in the place can be appropriate to take less. The allowable error of the model shape size varies with the model size, while the angle deviation is not affected by the model size. For the 4 m × 3 m magnitude of the low-speed wind tunnel test model of the sample plate and its parts of the machining accuracy and surface roughness is determined according to the following requirements. For the rest of the wind tunnel, the design tolerances of the model components can be appropriately scaled according to the data determined by the above principles. For example, for the wind tunnel of 8 m × 6 m scale, the data other than the angular tolerance can be scaled by 3 times of the root. For the standard calibration model, the design tolerance of each component should be slightly higher than the data provided above. The design tolerance of each part of the model is shown in Table 4.1. The requirements of surface roughness of each part of the model are shown in Table 4.2.

To ensure the required dimensional accuracy, wind tunnel test models are usually machined by high-precision CNC machine tools [2]. The above requirements for design tolerances, surface roughness, etc. are also applicable to wind tunnel test models Manufactured by additive manufacturing technology.

W. Zhu and D. Li, *Models for Wind Tunnel Tests Based on Additive Manufacturing Technology*, Additive Manufacturing Technology,
https://doi.org/10.1007/978-981-99-5877-1_4

Table 4.1 Design tolerances for each component of the model

Part name	Processing parameters	Tolerance requirements
Wing	Maximum clearance between the sample and the wing surface before 1/3 chord line	≤ 0.08 mm
	Maximum clearance between the sample and the wing surface in other parts	≤ 0.10 mm
	Maximum clearance between the surface at the wing isoparametric line and the knife ruler	≤ 0.10 mm
	Profile chord length	± 0.25 mm
	Exhibition length	± 1.50 mm
	Swept-back angle	$\pm 3'$
	Upper reverse angle	$\pm 3'$
	Turning angle	$\pm 3'$
	Trailing edge thickness	≤ 0.20 mm
Fuselage	Maximum clearance between the sample and the head surface of the body	≤ 0.15 mm
	Maximum clearance between the sample and other parts of the surface of the body	≤ 0.20 mm
	Diameter	± 0.25 mm
	Length	± 1.50 mm
Tailpiece	Maximum clearance between sample and tailplane surface	≤ 0.10 mm
	Maximum clearance between the tail surface at the equivalence line and the knife ruler	≤ 0.10 mm
	Profile chord length	± 0.20 mm
	Exhibition length	± 1.00 mm
	Swept-back angle	$\pm 3'$
	Upper reverse angle	$\pm 3'$
	Trailing edge thickness	≤ 0.20 mm
Externals	Maximum clearance between the sample and the surface of the external hangings	≤ 0.15 mm
	Length	± 0.50 mm
	Cross-sectional diameter	± 0.25 mm

Table 4.2 Surface roughness requirements for each part of the model

Model parts	Roughness requirements
Wing, tail, flaps, ailerons, rudder surfaces	0.1 mm
Fuselage, external object surface	0.2 mm
Fuselage and wing, fuselage and tail mating surface	0.1 mm
Standard calibration model surface	1.25 μm

4.1.2 Strength and Stiffness Calibration Requirements

The model design must be calibrated for strength and stiffness.

The safety factor for the strength check calculation is taken as 3, mainly checking the roots of the wing and tail, the pivot and fixed parts of the maneuvering surface and the connecting parts of the front and rear pivot points of the model. Especially for the wings with large span ratio and the wings tested under higher wind speed, the strength problem is most prominent [1].

The requirement for stiffness check is to add static maximum aerodynamic load to at least 5 profiles based on the exposed wing half-wingspan, with the load distribution according to GJB67.2-85 "Strength and Stiffness Specification for Military Aircraft", and calculate the resulting airfoil deformation with the following limits [2]:

(1) Low-speed wind tunnel test model: the distance between the deflection of the wing tip and the horizontal reference plane of the wing tip is not more than 15 mm (for the 3 m magnitude of the low-speed wind tunnel test model);
(2) High-speed wind tunnel test model: ① wing surface wing tip and wing root relative deflection: delta wing not more than 0.1°, swept-back wing not more than 0.3°; ② deformation of the upper (lower) counter angle of the wing surface: delta wing not more than 0.2°, swept-back wing not more than 0.5°; ③ deformation of the rudder deflection: delta wing and swept-back wing are not more than 0.1°.

Under high speed wind tunnel high speed pressure experimental conditions, the deformation of the model often exceeds the above limits. In order to obtain accurate experimental data, the deformation of the model is usually measured by optical instruments and the experimental data are corrected. Some wind tunnels measure (or calculate) the model deformation to improve the accuracy of the experimental data, even when the above deformation limits are not exceeded, and correct the experimental data.

4.2 Evaluation of Model Manufacturing Accuracy

4.2.1 Surface Roughness Analysis

Surface roughness is an important indicator to measure the surface processing accuracy of the parts, the surface roughness of the parts will affect the two matching parts have contact surface friction, wear and tear of the moving surface, the sealing of the fitting surface, the working accuracy of the mating surface, the fatigue strength of the rotating parts, the beauty of the parts, etc., and even the corrosion resistance of the parts surface have an impact.

The effect of the surface roughness of the wind tunnel test model on the experimental data is related to the type and size of the roughness and the local boundary layer state of the model [2]. In general, when the height of the roughness is less than the thickness of the bottom layer of the local turbulent boundary layer laminar flow, the roughness has no effect on the experimental data, i.e., the experimental data measured at such a roughness is the same as that measured on the smooth model surface. On the contrary, it has an effect. In general, the surface roughness mainly affects the model resistance; at sub and supersonic velocities, it mainly affects the frictional resistance and differential pressure resistance of the model; at transonic velocities, there are excitation waves on the model surface, and the mutual interference between the excitation waves and the boundary layer leads to a significant effect of the rent roughness on the wave resistance. When there is strong interference between the surge and the boundary layer and airflow separation occurs, the magnitude of roughness also changes the model pressure distribution, which has an effect on lift and moment. When large head-on angle experiments are performed, the size of the rent roughness changes the head-on angle at which the model appears asymmetric vortices, leading to an effect on the cross-side aerodynamic characteristics from the surface.

This book uses the surface roughness meter TR240 to measure the surface roughness of resin model before and after regrinding and the surface roughness of electrodeposited copper-nickel, the roughness measurement curve schematic diagram is shown in Fig. 4.1. The average values and the corresponding deviations were obtained for the unground resin parts, ground resin parts, electroplated parts with copper deposition and electroplated parts with nickel deposition, as shown in Fig. 4.2. The surface roughness of the resin part is larger than that of the copper deposited part due to the step effect, while the roughness of the reground resin and electrodeposited part is very small, and the surface roughness of the nickel deposited part is better than that of the copper deposited part due to the extremely fine crystallization of nickel, with a Ra value of 1.48 μm. It can be seen that the simplicity and convenience of resin model surface treatment is particularly suitable for wind tunnel test model fabrication where only aerodynamic profile surface is required.

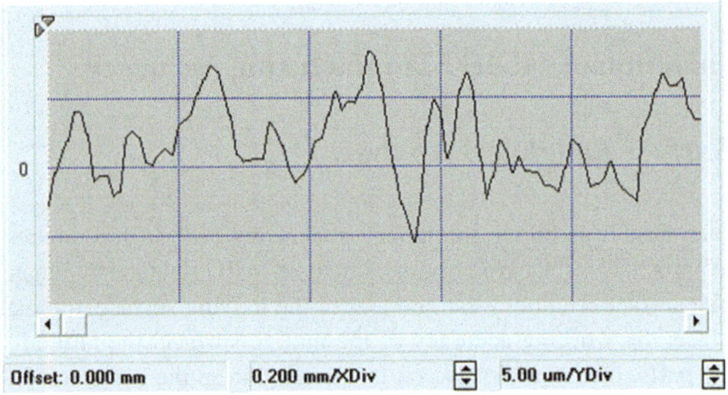

Fig. 4.1 Schematic diagram of roughness measurement curve

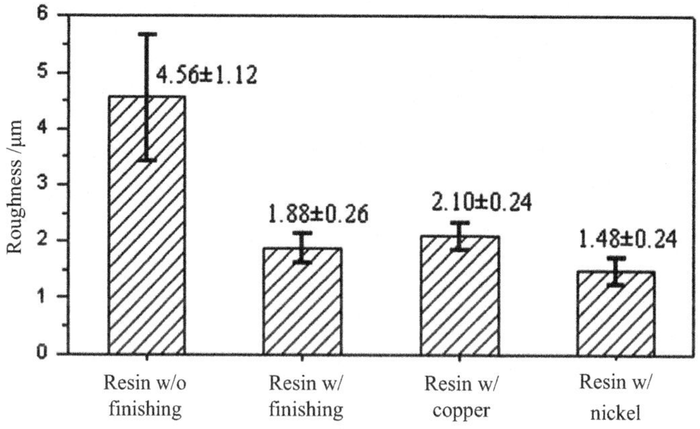

Fig. 4.2 Comparison chart of roughness after various surface treatments

4.2.2 *Manufacturing Accuracy Analysis*

In order to obtain the influence of 3D printing process and electrodeposition process on the surface shape, a surface shape is designed as an accuracy test piece, and this model is used to analyze the influence of the manufacturing process on the accuracy of the surface shape, and to propose further manufacturing optimization methods for the 3D printing of the wind tunnel test model.

As Fig. 4.3 shown, the obvious step effect is visible through the 3D printing process. After the resin model was resharpened to form a smooth surface, and then the resin model was shielded and protected in all planes, and only the surface of the surface was given electrodeposited nickel. The surface data of the model was obtained through the LSH 3D laser measurement system (measurement accuracy: $\pm$ 0.05 mm), and the resharpened resin surface and the surface after electrodeposition were measured separately. The point cloud data was processed in CATIA V5 software. The point cloud was imported in the Digitized Shape Editor module and the cluttered and useless points were removed. Since the coordinate system where the point cloud is measured does not correspond to the coordinate system of the theoretical model, another axis system is constructed under the measurement coordinate system so that the point cloud data can be relocated to match the location of the theoretical surface. The surface point cloud is separated from each plane point cloud, and each plane point cloud is extracted by activating the Basic Surface Reconstruction command in the Quick Surface Reconstruction module to construct each datum. This allows you to optimize the composition of the datum by repeatedly building two perpendicular planes. Finally, the data from the measurement coordinate system is converted to the theoretical model coordinate system using Axis To Axis, and the point cloud is migrated to the spatial location where the comparison can be performed.

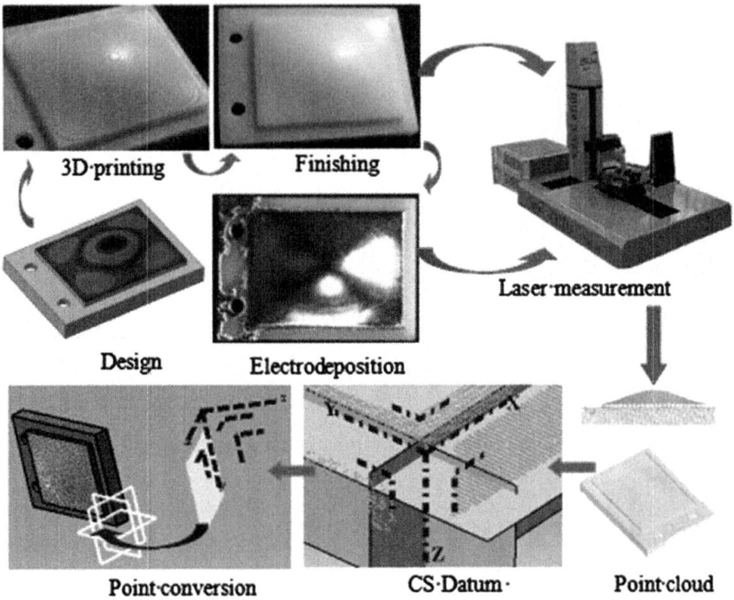

Fig. 4.3 Flow chart of manufacturing and measurement of surface accuracy test parts

The distance analysis of the point cloud measurement data before and after the resin electrodeposition and the theoretical surface, the overall error distribution is shown in Fig. 4.4. From the error distribution cloud, the distance thickness at different locations can be visualized, so the 3D measurement can be used as a calibration method for the post-processing of the resin wind tunnel test model. The uniformity of polishing is detected by inverse measurement, so that the thinner areas can be marked for re-polishing. The amount of polishing should be small rather than large so as not to over polish some areas and make it difficult to coordinate the overall repair. Figure 4.5 shows the results of the percentage distribution of the different error ranges. As can be seen from the figure, 96.64% of the point clouds were in the error range of − 0.20 to 0 mm in the surface post-processing, and 98.92% of the point clouds were in the error range of 0 to 0.20 mm in the surface electrodeposition. Considering that the characterization processing accuracy of the 3D printer used is ± 0.1 mm, it can be seen from the figure that the different processing steps keep the accuracy of the model surface well. From the above analysis, we can indirectly get the removal amount of step effect is about − 0.1 mm and the thickness of plating is about + 0.2 mm. Therefore, in the offset compensation of model design, not only the reserved thickness of electrodeposition should be considered, but also the removal amount of step effect and surface finishing of 3D printed parts must be considered, and the processing amount of different parts will follow the relevant process for guidance.

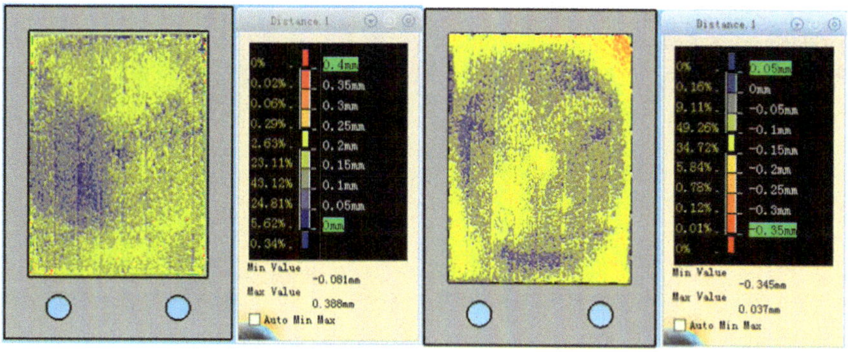

(a) (b)

Fig. 4.4 Overall error distribution before and after electrodeposition. **a** Error cloud of regrinding resin surface, **b** Error cloud of electrodeposited surface

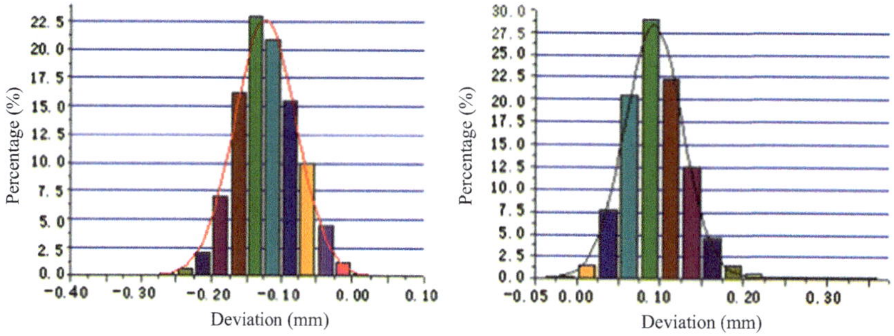

(a) Statistical distribution of errors on regrinded resin surface

(b) Statistical distribution of errors on electrodeposited surface

Fig. 4.5 Statistical histogram of the error range before and after electrodeposition

The model shown in Fig. 4.6 is divided into 19 profiles in the horizontal direction, marked as −9~0~9, and the point cloud data on each profile is obtained by using Planar Sections in the Digitized Shape Editor module, and the points are successively converted into lines by Curve From Scans to obtain the profile lines of the resin surface and the nickel deposition surface. In the longitudinal direction, 19 planes were also taken to intersect the curve into a finite number of points, and the average error between the surface points of resin and the surface points of nickel layer at each profile was found and the theoretical profile line, respectively. Similarly, the profile contour lines on the midline L1 and diagonal L2 and L3 of the model are extracted, and the thickness uniformity of the resin electrodeposition can be specifically analyzed.

In Fig. 4.7, the comparison of the error before and after electrodeposition of each profile of the test model shows that the polishing volume is larger on both sides of the model when resin polishing, because the curvature changes slowly on both sides

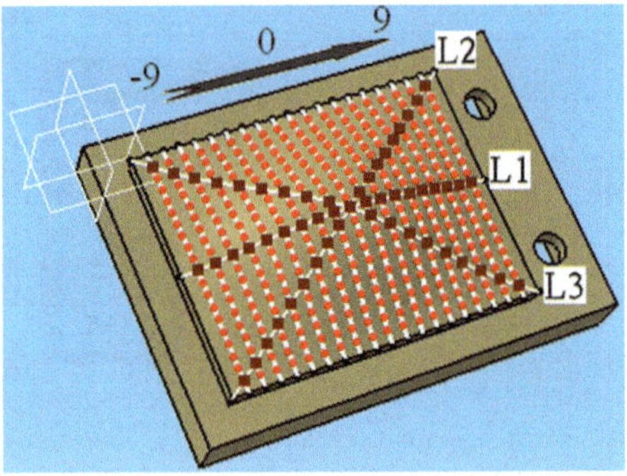

Fig. 4.6 Point-line surface position markers for error analysis

and the volume error of the step effect is larger, and the polishing volume here is increased accordingly. In contrast, the error formed by the measured point cloud of electrodeposition compared with the theoretical surface is not as obvious on both sides of the model as when resin polishing, which makes up for the missing part of over-polishing to a certain extent, indicating that a thicker deposited layer is formed on both sides of the model. It can be seen that the amount of resin polishing and the uniformity of electrodeposition compensate each other to ensure the accuracy of the electrodeposited surface.

Electrodeposition is a process in which the positive and negative ions in the plating solution move in a directional manner along the electric power lines under the action of external electric field. The ability to make the deposited layer of metal deposited uniformly on the uneven surface of the model is called the dispersion ability of the plating solution. The factors affecting the dispersion ability are mainly geometric and electrochemical factors. Geometric factors mainly refer to the shape of

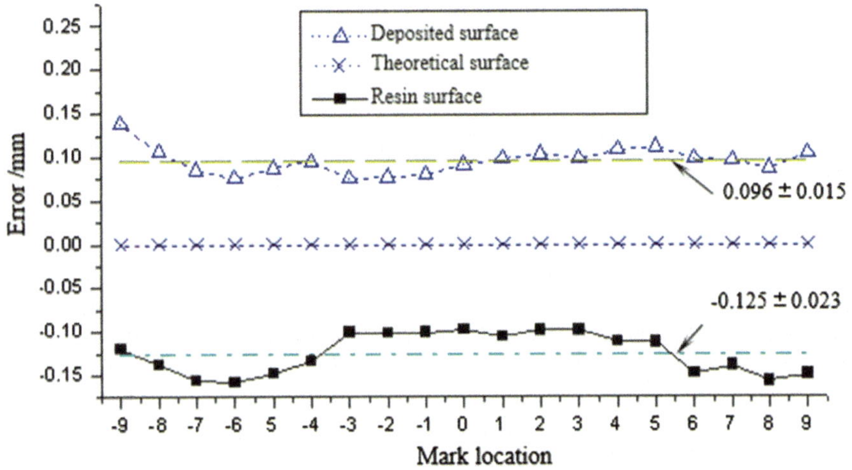

Fig. 4.7 Comparison of errors before and after electrodeposition for each profile of the test model

the plating bath, the shape of the anode, the shape of the part and the mutual position and distance between the part and the anode. The elements are tip discharge, edge effect, distance of cathode and anode, etc. It mainly affects the current distribution of plating. Electrochemical factors include polarization, current density, conductivity of the solution, and current efficiency, etc.

The Z-directional absolute coordinate values of the points on lines L1, L2, and L3 were extracted to construct the contour lines of the test surface, and the distance between the resin model and the electrodeposition model was enlarged by the coordinate system, as Fig. 4.8. The comparison of the deposited layers of L1, L2, and L3 are shown respectively. It can be seen that the deposition thickness at the top of the model is larger, because the surface of the model faces the anode plate during electrodeposition, and the top surface is closer to the electrode which makes the power lines shorter and concentrated. The main surface of the plated part should face the anode and be parallel to it, and the distance between different parts of the cathode and the anode should be reduced, and if necessary, pictorial anode and auxiliary anode should be used to ensure the uniform distribution of power lines. In a certain range, increasing the distance between cathode and anode can improve the dispersion ability, and theoretically the power line is the best when the plated part is at infinity. However, in reality, the distance between the electrodes cannot be increased indefinitely, otherwise the width and floor space of the plating tank will increase, and the tank voltage will increase and the energy consumption will be high.

The distribution of power lines in the end corner area will be significantly higher, from the figure can also be seen on both sides of the model at the edge of the thicker plating, especially L2, L3 edge thickness than L1 obvious, because L2, L3 are in the intersection of two right-angle edge, while L1 only a right-angle edge. Therefore, to prevent the concentration of power lines to produce "edge effect", the layout of the anode should be dense in the middle, sparse on both sides, the two ends of the slot to leave a certain empty space, while the cathode rod to hang full. Metal protection wire can be hung at the tip of the part to disperse the current on the cathode, this metal protection wire is called the protection cathode.

To obtain a uniform coating, in addition to choosing a reasonable solution composition and improving the formula, reasonable operation, mounting parts and taking some special measures are very important in production, usually using the following methods [40]:

(1) Impulse current: Use a current 2–3 times higher than the normal current for a short time to impinge on the plating at the beginning of plating.
(2) reasonable hanging parts: so that the parts in the best current distribution state, while not making the precipitation of gas stagnation in the blind holes, low-lying parts of the parts.
(3) Adjust the distance between the cathode and anode according to the possibility, and reduce the distance ratio between different parts of the cathode (i.e., the concave and convex parts of the part) and the anode.
(4) Improve current distribution by using "pictorial" anodes.

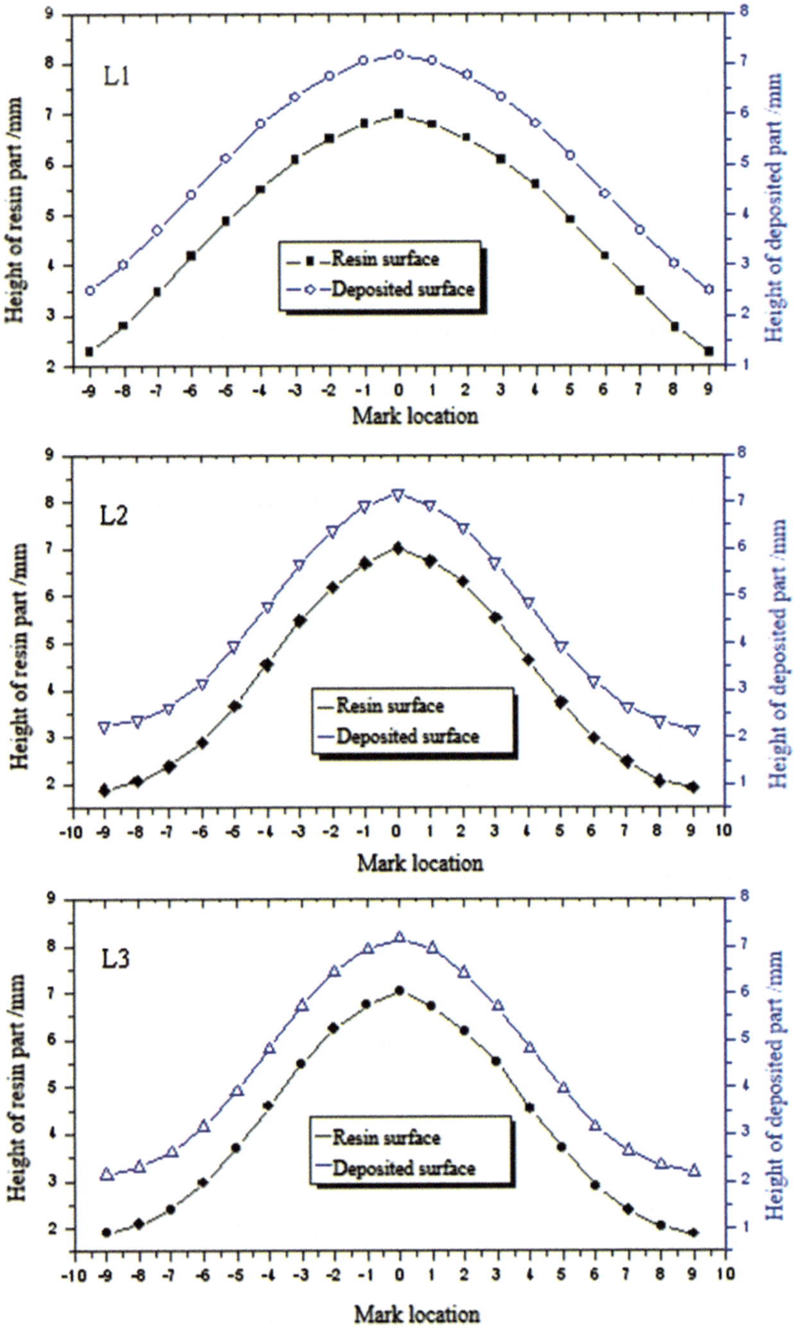

Fig. 4.8 Comparison of profile lines L1, L2 and L3 before and after electrodeposition of the test model

(5) The use of protective cathodes and shielding protection to reduce the current density in concentrated areas of power lines.

4.3 Evaluation of the Mechanical Properties of Models

4.3.1 Mechanical Properties Testing of Model Materials

According to the ISO plastic tensile and bending test standard [41, 42], the relevant mechanical experiments were performed and the tensile and bending specimens were manufactured as shown in Fig. 4.9. The cross-section of the 3D printed tensile specimen is 10 × 4 mm and the cross-section of the bending specimen is 15 × 6 mm. Different thicknesses were electrodeposited on them, and the properties of the resin-metal material with different thicknesses were obtained by experiments as shown in Figs. 4.10 and 4.11. The properties of the resin-metal materials with different deposited thicknesses are shown in Figs. 4.10 and 4.11. The theoretical values of Young's modulus of the composites in the figures were calculated using the prediction equation [43].

$$E_L = E_f V_f + E_m(1 - V_f) \tag{4.1}$$

Eq

E_L modulus of elasticity of the composite;
E_f the modulus of elasticity of the deposited metal;
V_f volume percentage of metallic material;
E_m modulus of elasticity of the resin matrix.

As seen in the figure, the stiffness and strength of the composite increased with the increase of the deposition thickness. The stiffness strength of the nickel deposition

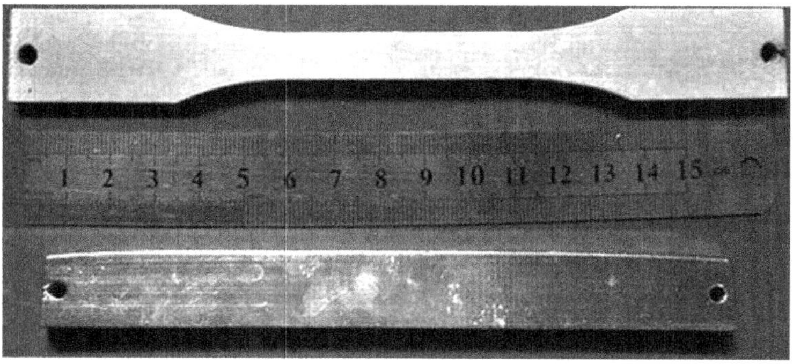

Fig. 4.9 ISO standard tensile and bending test specimens

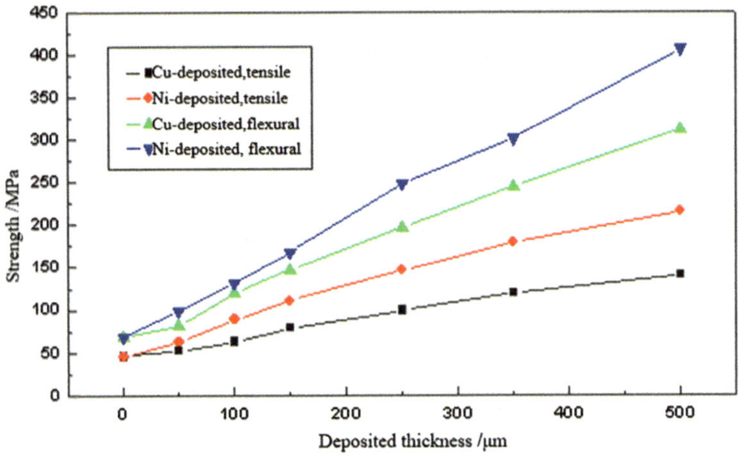

Fig. 4.10 Strength of resin copper and nickel electrodeposited composites

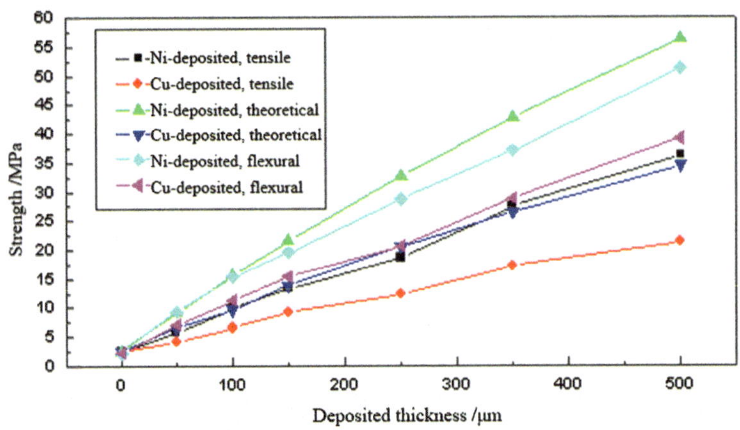

Fig. 4.11 Stiffness of resin copper and nickel electrodeposited composites

is greater than that of the copper deposition for the same deposition thickness. The actual Young's modulus of the copper-nickel composite is smaller than the theoretical calculated value due to the bond strength at the resin-metal deposition interface. Since the accuracy of electrodeposition decreases with the increase of deposition thickness, the deposition thickness is taken as 0.1–0.2 mm. Comparing the original 14120 resin with SL-Ni-0.1 with nickel deposited thickness of 0.1 mm, the tensile strength increased from 45.7 to 89 MPa, the elastic modulus increased from 2.5 to 10 GPa, the bending strength increased from 68.9 to 131.6 MPa, and the bending modulus increased from 2.3 to 15.4 GPa.

4.3.2 Model Numerical Analysis of the Mechanical Properties

1) The choice of analysis method and analysis software

Designers often use CFD tools to perform various virtual tests, quickly conduct comparative parametric studies and multiple solution screening, so that only a small number of design solutions that do have practical applications need to be wind-tunnel tested. Wind tunnel tests and CFD methods are two complementary tools in aircraft aerodynamic layout design, and the comparison of calculation accuracy with wind tunnel tests and test flight results is generally regarded as private property by each aircraft company and kept confidential.

The CFD method mainly analyzes the aerodynamic performance of the aircraft shape, while for the actual aircraft model that needs to be manufactured, there exists a requirement for mechanical properties, i.e., for strength and stiffness calibration. This requires obtaining the surface pressure flow field of the model from CFD and analyzing the strength and stiffness of the wind tunnel test model for different materials and structures under this pressure load. Considering the possible damage in the strength of the 3D printed based wind tunnel test model in the dangerous sections such as the root of the wing and tail, and the joint part of the maneuvering surface deflection, the finite element method is proposed to be used to calibrate the wind tunnel test model. The finite element analysis tool can predict the wind tunnel test results exactly, so as to check the validity of the wind tunnel test model manufacturing, prevent the model from affecting the test accuracy due to the abnormal performance during the blowing wind, and avoid the growth of the test cycle and the increase of the test cost caused by the rework of the model.

The selection of loading method is one of the important steps of engineering analysis using finite elements. The choice of different loading methods will produce different analysis results and thus affect the validity of the analysis results. Commonly used approximate loading methods are [44]:

(1) Compression core concentrated force loading: the loading method is simple, but its results do not reflect the location of the real danger point and the maximum stress value, and cannot correctly reflect the force and deformation of the structure.
(2) Split-plane force loading: theoretically, split-plane force loading is the most approximate simulation of load distribution, but it is more complicated in practice.
(3) Block concentration force loading: the results obtained are very close to those obtained by the block surface force loading method, and the operation is very simple.

With the development of finite element analysis technology, the analysis of the force deformation of solids in the flow field can be realized more realistically by using the method of fluid–solid coupling. The existing fluid–solid coupling software is used in the form of ADINA, ANSYS + CFX, FLUENT + MPCCI + ABAQUS, etc. ADINA has a fluid–solid coupling (FSI) solution function that can simulate in a single system ADINA the physical phenomenon of complete coupling between a fluid and a structure that experiences significant nonlinear response due to large deformation, inelasticity, contact and temperature.. A fully coupled fluid–structure coupling model means that the deformation of the structure affects the fluid region and in turn the forces of the fluid are applied to the structure [45].

From a fluid perspective, Navier–Stokes fluids can be non-compressible, slightly compressible, low and high velocity compressible [46]. From the structural point of view, various structural cell types can be involved in the FSI process supporting various material models, supporting various nonlinear physical processes such as material failure, cell life and death, structural instability, phase transitions, etc. ADINA combines the structural and fluid dynamics equations in a single system to obtain a unified set of equations for this system and to solve it. ADINA-FSI is unique in that because it offers two different methods, direct FSI coupling and iterative FSI coupling, to solve the coupling between the fluid model and the structural model. In both cases, displacement consistency and force balance conditions are satisfied at the fluid–structure coupling interface.

2) Specific steps of fluid–solid coupling analysis

The parameters used to build the fluid–solid model in ADINA software are shown in Table 4.3 where the Poisson's ratio of the nickel-deposited material is obtained from the composite Poisson's ratio prediction equation at [43]

$$\nu_L = \nu_f V_f + \nu_m \left(1 - V_f\right) \tag{4.2}$$

Eq

ν_L Poisson's ratio of the composite;
ν_f Poisson's ratio of the deposited metal;
V_f volume percentage of metallic material;
ν_m Poisson's ratio of the resin matrix.

The specific setup steps for ADINA fluid–solid coupling analysis are as follows [47]:

(1) Format conversion: Convert the wing model in CATIA to *.x_t format.
(2) Establishment of solid model (ADINA-Structure)

   a. Read in the model in wing x_t format.
   b. Define the analysis type: select static analysis and fluid–solid analysis.
   c. Set the material: isotropic elastic material, enter the modulus of elasticity and Poisson's ratio.
   d. Define the boundary conditions: constrain the degrees of freedom of the assembly end faces, no loading is required for the model, and the forces are transferred by the flow field.

Table 4.3 Table of parameters used in the fluid–solid model

(a) Solid-wing (angle of attack α: 20°)

Materials	Modulus of elasticity E/GPa	Poisson's ratio ν
Somos14120	2.3	0.23
SL-Ni-0.10	10	0.24
SL-Ni-0.15	13	0.24
SL-Ni-0.25	19	0.24
Al alloy	70	0.30

(b) Fluid-air

Mach number *Ma*	Dynamic viscosity μ/Pa S	Density ρ/kg M^{-3}	Bulk modulus *K*/Pa
0.2–0.8	1.8×10^{-5}	1.23	1.4×10^{5}

e. Define the cell group type: select the solid cell type 3-D Solid.
f. Define the fluid–solid boundary: select the wing surface as the FSI boundary.
g. Specify the cell size: Set the cell size to 0.01.
h. Delineate the mesh: generate a solid four-node cell mesh, the flow-solid boundary will be highlighted and bolded, as in Fig. 4.12a shows.
i. Solution control: Set the maximum number of iterations to 50 and the error of iteration to 0.005.
j. Save the file *.idb and output the solved data *.dat.

(3) Fluid modeling (ADINA-CFD)

a. The flow field x_t model can be read in, as well as the wing model, and then a new body is created in ADINA-M and generated by subtracting the Boolean operation.
b. Define the analysis type: select 3D problem, turbulence analysis, slightly compressible, no heat transfer.
c. Set the fluid parameters of the air: define the kinetic viscosity, density, bulk modulus and select the standard K-Epsilon parameters.
d. Define the pressure boundary condition: Define the boundary condition with relative pressure = 0. Apply this constraint to the other five external boundaries except for the incoming flow port.
e. Define the incoming flow velocity: different wind speeds are applied perpendicular to the inlet boundary surface.
f. Define the fluid–solid coupling boundary condition: choose the wing surface as the fluid–solid coupling boundary.
g. Define the cell group: select the entity cell type 3-D Fluid.
h. Specify the grid density: specify the grid size around the flow field as 0.1, and specify the grid size at the flow-solid boundary as 0.01.
i. Delineate the mesh: generate the solid four-node cell mesh, as Fig. 4.12b shows.

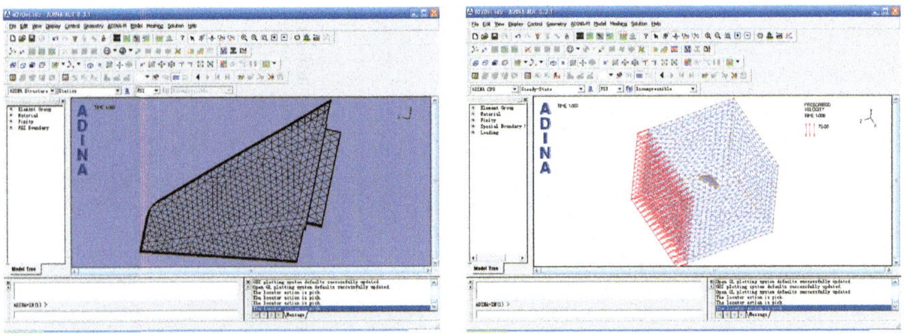

(a) Airfoil solid finite element model (b) Air fluid finite element model

Fig. 4.12 Modeling of solids and fluids in fluid–solid coupling analysis

j. Delineate the flow-solid boundary layer grid: divided into two layers, the thickness of the first layer is 0.005, and the total layer thickness is 0.02.
k. Solution control: No FBCI algorithm is used, and the maximum number of iterations is set to 50, with an iteration error of 0.005.
l. Save the file *.idb and output the solved data *.dat.

(4) Start ADINA-FSI, call in the solids and fluids solution dat file, and set the appropriate memory size for the calculation.
(5) Enter the *.por file post-processing and extract the analysis results.

3) Predictive analysis of the rigidity strength of fluid–solid coupling

The size of the model is limited by the boundary of the test section of the wind tunnel and cannot exceed a certain range, otherwise the boundary has a large impact and it is difficult to make accurate corrections; for the three-dimensional large span ratio aircraft model, the ratio of the wing span b to the test section width w is generally not greater than 0.6 for the high-speed wind tunnel and not greater than 0.7 for the low-speed wind tunnel. For the small span ratio aircraft model, depending on the experimental angle, the general requirement is $b/w \leq 0.5$ [1]. Taking the maximum value of the ratio of wing span to test section width 0.7, the distance standard of wing tip deflection in the low-speed wind tunnel of 3 m magnitude is converted to a smaller angle standard of 0.818°. The wing model designed in this book has a half wing span of 210 mm, then the wing tip deflection cannot exceed 3.00 mm.

The maximum deformation DMX of resin, nickel-deposited resin, and aluminum alloy at a wind speed of 100 m/s is shown in Table 4.4. By comparing the relative deflection angles of wing tip and wing root, it can be seen that the resin material does not meet the low speed requirement, while SL-Ni-0.1 and aluminum alloy both meet the requirement, and the deformation of aluminum alloy is very small, and the analyzed flow field pressure diagram and wing deformation diagram are shown in Figs. 4.13 and 4.14. Therefore, in terms of stiffness requirements, the pure resin material cannot be used for the wind tunnel test model with a large span-to-chord ratio, and the strength of the model can be improved and the stiffness performance can be improved to some extent by depositing metal on the resin surface or by lining with metal plates. The following is an analysis of the applicable wind speed range only for composites with different thicknesses of nickel layers deposited.

At the angle of attack of 20°, the maximum stress and maximum deformation of Mach number from 0.2 to 0.8 were analyzed, and the stress distribution schematic

Table 4.4 Comparison of wing tip deformation for different materials ($Ma = 0.3$)

Materials	Maximum deformation DMX/mm	Arctan(DMX/210)/°	Low-speed wing tip deflection conversion values
Somos14120	3.80	1.037	$\text{Arctan}\left(15 \Big/ \left(\frac{1}{2} \times 3000 \times 0.7\right)\right)$
SL-Ni-0.1	0.88	0.240	
Al alloy	0.12	0.032	$= \text{Arctan}(\mathbf{3.00}/210) = 0.818°$

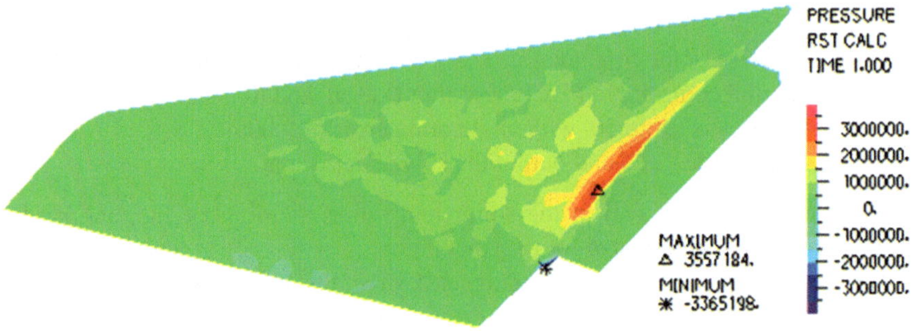

Fig. 4.13 Airfoil surface pressure diagram

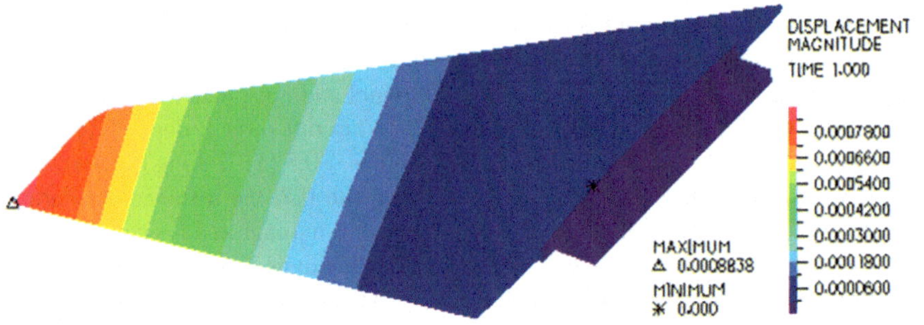

Fig. 4.14 Wing deformation distribution

is shown in Fig. 4.15. The maximum stress increased from 3.5 to 66.7 MPa, and the maximum deformation increased from 0.43 to 4.42 mm, the Figs. 4.16, 4.17 and 4.18 show the maximum stress and maximum deformation values for depositing 0.1 mm, 0.15 mm and 0.25 mm nickel layers, respectively. The available blowing data are summarized in Table 4.5, that is, if the safety factor is 3, the tensile strength of SL-Ni-0.10 is 89 MPa to meet the 0.5 Ma requirement; the tensile strength of SL-Ni-0.15 is 111 MPa to meet the 0.6 Ma requirement; and the tensile strength of SL-Ni-0.25 is 147 MPa to meet the 0.7 Ma requirement. And the wing tip deformation at these Mach numbers is within the range of 3.00 mm, and the materials of different thicknesses meet the requirements of both strength and stiffness at the permissible Mach numbers. Therefore, the applicable wind speed range of resin-nickel deposited materials can cross from low speed to subsonic speed. The higher the wind tunnel test wind speed, the higher the model accuracy requirements and stiffness strength requirements. The larger the deposition thickness, the higher the stiffness and strength, but the lower the deposition accuracy. Therefore, the deposition thickness should be selected to meet the stiffness and strength and also to ensure the processing accuracy.

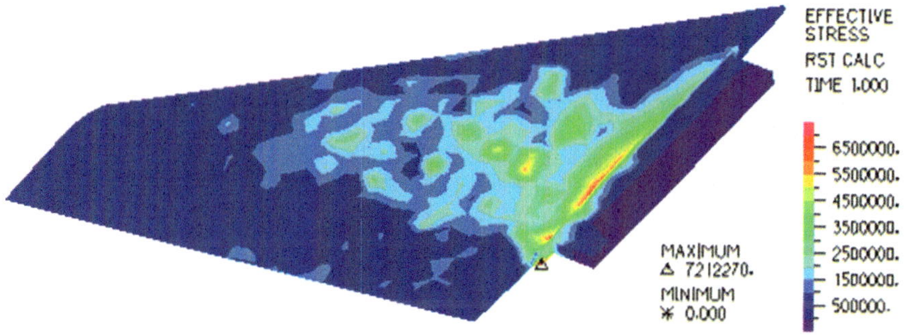

Fig. 4.15 Airfoil surface stress diagram

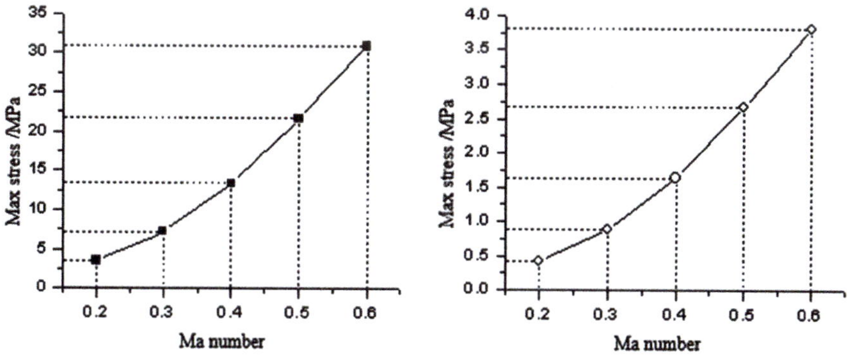

Fig. 4.16 SL-Ni-0.10 maximum stress and maximum deformation values

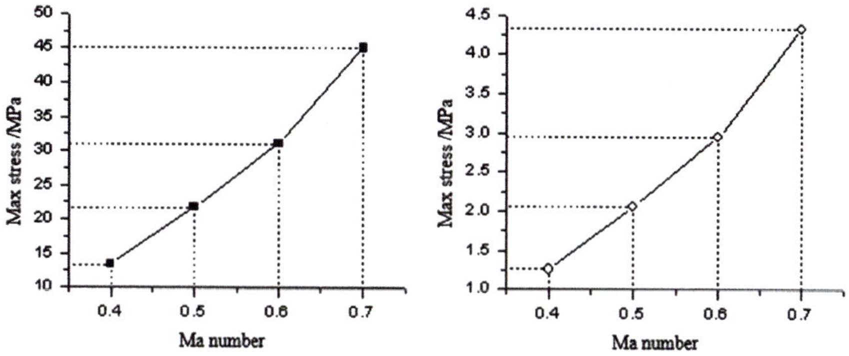

Fig. 4.17 SL-Ni-0.15 maximum stress and maximum deformation values

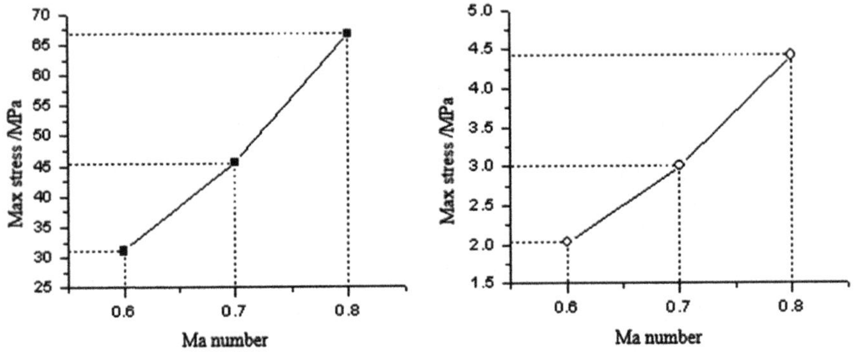

Fig. 4.18 SL-Ni-0.25 maximum stress and maximum deformation values

Table 4.5 Permissible Mach numbers for different deposition thicknesses

Nickel deposit thickness/mm	Tensile strength σb/Mpa	Permissible stress/MPa I Permissible deformation/mm (Safety factor: 3)	Permissible Mach number *Ma* (Maximum stress/MPa I Maximum deformation/mm)
0.10	89	29 I 3.00	0.5 (21.7 I 2.67)
0.15	111	37 I 3.00	0.6 (30.9 I 2.95)
0.25	147	49 I 3.00	0.7 (45.5 I 3.00)

When the Mach number is 0.8–1.4, it has entered the transonic range. Transonic region from the aircraft surface at a point of the so-called critical speed of sound speed to the entire flow field are supersonic until, is the aircraft surface of the airflow both subsonic and supersonic mixed flow area. When the aircraft reaches the critical speed, a surge is formed on the surface and develops as the Mach number increases. The surge generates wave resistance, which increases the drag force several times compared to subsonic velocity. As a result, the lift is reduced, the center of pressure is shifted back, the moment changes abruptly, and the aircraft may vibrate or chatter. Various instruments will be shaken by the excitation wave. The measures to overcome the adverse effects of transonic velocity are the use of small span ratio, small thickness ratio swept back wing and research on supercritical wing and fuselage shaping according to the area law, etc. The composite materials used in the transonic wind tunnel test model are to be studied later.

4.3.3 Reliability Testing of Plastic Model Parts

The full-mode force measurement test is performed by using a six-component balance to measure the aerodynamic forces of the model at a range of attitude angles under a given dynamic pressure condition. To perform the full-mode force measurement test, all subsystems of the wind tunnel are operated in a coordinated manner to keep the test dynamic pressure stable and to change the model attitude automatically according to the test requirements to achieve accurate and efficient measurement of the model aerodynamic force.

In the force measurement test, when the side slip angle of the model is 0°, the measurement under a series of head-on angles is called longitudinal force measurement test; given a certain head-on angle of the model, the measurement under a series of side slip angles is called transverse force measurement test; given a certain non-zero side slip angle of the model, the measurement under a series of head-on angles is called quasi-longitudinal force measurement test; under different dynamic pressure conditions, the force measurement of the same model state is called variable Reynolds number test; Multiple repeated measurements of the same model state under the same test conditions are called repeatability tests [1]. In order to deduct the bracket interference, the bracket interference test is also performed.

In a low-speed wind tunnel force measurement model for an aircraft design, the fuselage is machined in metal and the wing and tail are manufactured by 3D printing. The wind tunnel test was conducted in the FL-12 low-speed wind tunnel, and the maximum wind speed was up to 100 m/s. By changing the angle of attack to blow the wind test and collecting the test data, when the angle of attack was −20° and the wind speed accelerated to nearly 70 m/s, the vertical tailplane broke from the root of the connection, as shown in Fig. 4.19. The possible causes of this fracture were insufficient strength of the resin model on the one hand, and design or manufacturing defects of the wind tunnel test model on the other hand. Since the much weaker parts of the wing such as the front flap and ailerons were not abnormal, the main cause of the tail fracture should originate from the structural defects of the model. For example, the interlayer defects introduced during the layer-by-layer 3D printing of the tailplane may be the main cause of this fracture. On the one hand, the designers should choose the printing direction of the model parts reasonably in the main bearing direction to avoid making the model print the maximum load in the interlayer direction; on the other hand, the process personnel should reduce the interlayer defects by choosing stable raw materials and equipment, optimizing the printing process parameters and other measures.

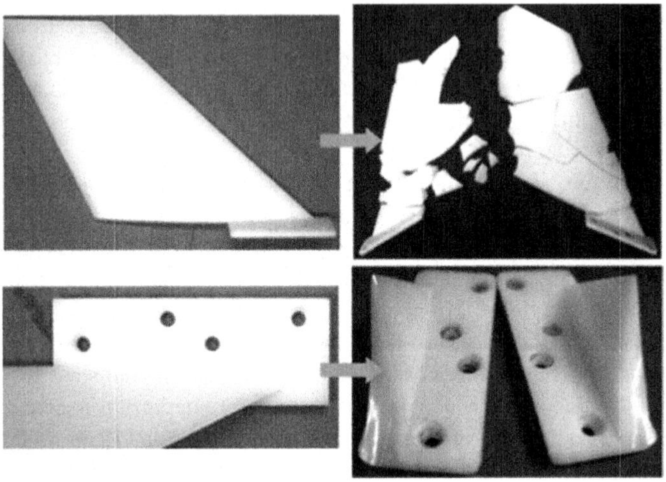

Fig. 4.19 Wind tunnel force measurement test results

Chapter 5
Electrodeposition Strengthening of Plastic Wind Tunnel Test Models

Many of the 3D printing materials used for model manufacturing are polymers, such as photosensitive resins (light-curing 3D printing) and thermoplastics (melt-extrusion 3D printing). Compared to metals, the mechanical properties of polymer materials are poor. For thin and long parts in wind tunnel test models, such as wings, flaps and ailerons, they cannot be strengthened by embedded metals and other means, and other strengthening techniques need to be studied. Zhou Zhihua et al. of Xi'an Jiaotong University proposed the method of electrodeposition of metals on the surface of resin models to improve the mechanical properties and serviceability of the models [29, 30]. In this chapter, the technology will be introduced, and the influence of resin surface electrodeposition process parameters on the surface quality of the model will be investigated through process experiments, and the influence of deposition thickness and bond strength of resin-metal interface on the mechanical properties of composite specimens will be analyzed through mechanical experimental studies, and the role and effect of chemical roughening on improving the bond strength of the interface will be investigated, and finally the economics of model manufacturing will be analyzed with an example of wing pressure measurement model electrodeposition manufacturing.

5.1 Experimental Study of Electrodeposition Process

Resin model surface electrodeposition process to include surface roughening, conductivity and electrodeposition, the process is as shown in Fig. 5.1.

W. Zhu and D. Li, *Models for Wind Tunnel Tests Based on Additive Manufacturing Technology*, Additive Manufacturing Technology,
https://doi.org/10.1007/978-981-99-5877-1_5

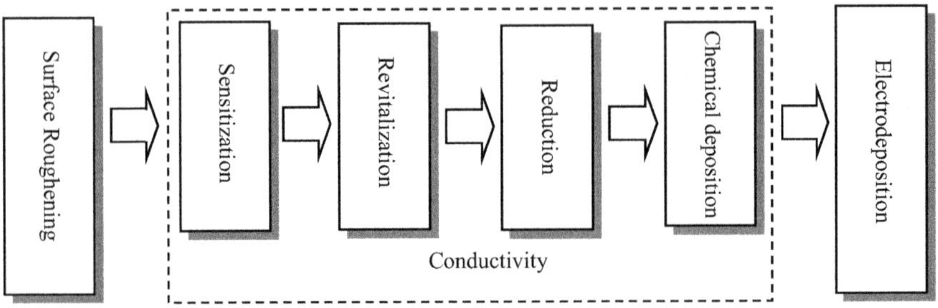

Fig. 5.1 Resin model surface electrodeposition process

5.1.1 Surface Roughening

The purpose of roughening is to make the resin surface microscopically rough, so that the contact area between the deposited layer and the substrate increases, and also to change the resin surface from hydrophobic to hydrophilic, in order to improve the bonding force between the resin surface and the deposited layer. The effect of roughening treatment has a large impact on the resin-metal interface bond strength and its mechanical properties. For the light-curing resin material and wind tunnel test model with complex shape, this book selects chemical roughening method as its surface roughening treatment method, which has the characteristics of simple composition, convenient maintenance, fast roughening speed and good effect.

5.1.2 Conductivity

Since the resin wind tunnel test models manufactured by light-curing 3D printing are insulators, they cannot be directly electrodeposited and need to be conductive for subsequent electrodeposition, which consists of sensitization, activation, reduction and chemical deposition processes. The wind tunnel test model requires sufficient bonding between the metal and the substrate as well as sufficient coverage, and chemical deposition is chosen as the conductive treatment method for the resin model in this book. The treatment process of chemical deposition conductivity includes cleaning, sensitization, activation, reduction and chemical deposition, which play a key role in the quality of the chemically deposited layer.

5.1.3 Electrodeposition

After the resin model has been chemically deposited for conductivity, the metal layer is thickened using the usual electrodeposition process. The electrodeposition cathode is the resin model part to be electrodeposited, and the anode is the metal material

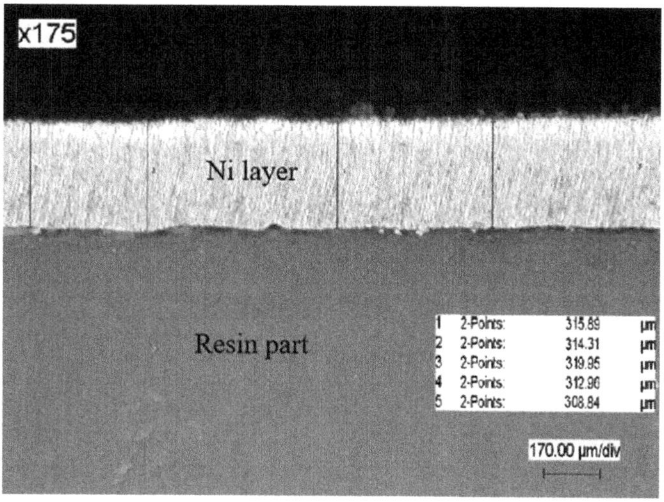

Fig. 5.2 Schematic diagram of nickel layer thickness measurement

plate used for electrodeposition. Nickel is chosen as the electrodeposition material for the resin surface because of its high strength, high stiffness, easy deposition and high hardness and wear resistance. Since this book uses electrodeposition method to enhance the resin wind tunnel test model, the deposition thickness is critical to the accuracy of the model, so it is necessary to quantify the deposition thickness according to the deposition time. According to Faraday's law of electrolysis, it is known that different deposition thicknesses are obtained for different deposition times at preset current density conditions. In order to determine the difference between the nickel deposition thickness calculation formula and the actual process, electrodeposition experiments were taken to verify it. The nickel electrodeposition layer thickness was measured by an optical microscope system on an intercepted cross-section and averaged over five measurements, as Fig. 5.2 is shown.

5.2 Surface Quality Analysis

Model surface quality has a large impact on wind tunnel experimental data, mainly related to the type and size of roughness and the state of the local boundary layer of the model [2]. Therefore, the surface quality of wind tunnel test models for nickel electrodeposition is mainly evaluated by surface roughness measurements and micromorphological observations. In this book, a portable surface roughness meter is used to measure the surface roughness of resin models before and after regrinding, as well as the surface roughness of electrodeposited nickel sulfamate and nickel sulfate.

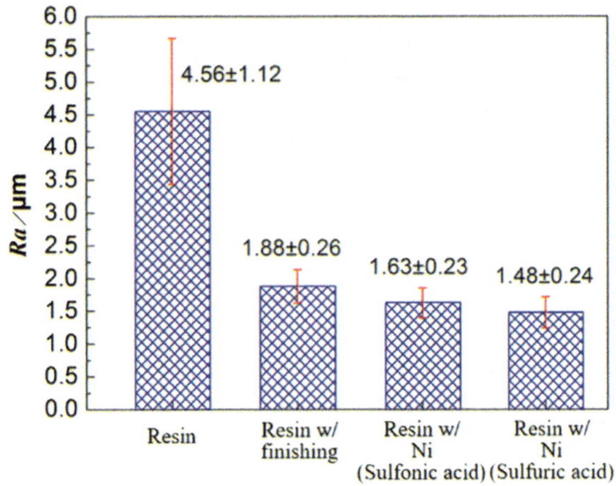

Fig. 5.3 Comparison of average surface roughness *Ra*

The average values and corresponding deviations were obtained for the unground resin parts, the resin parts after custom grinding, the nickel sulfate electrodeposited parts and the nickel sulfate electrodeposited parts, and their average surface roughness *Ra* were 4.56 μm, 1.88 μm, 1.63 μm and 1.48 μm, respectively, as shown in Fig. 5.3. As seen from the figure, the resin parts will have larger roughness as well as deviation due to the step effect, and the roughness of the reground resin and electrodeposited sample parts is smaller and related to the degree of regrinding. It was also found that the surface roughness of nickel sulfate deposition was better than that of nickel sulfamate, with a *Ra* value of 1.48 μm, while the roughness of resin models depended on the degree of regrinding relative to the surface roughness requirement of wind tunnel test models. It is particularly suitable for model manufacturing with high surface quality requirements. The surface quality for nickel deposition is not only related to the degree of regrinding of the resin substrate, but also to the microscopic surface state.

The microscopic surface of the electrodeposited nickel layer was observed using a scanning electron microscope, both at a magnification of 5000× as shown in Fig. 5.4. Scanning electron micrographs of nickel layer surfaces obtained by electrodeposition using nickel sulfamate formulation at different current densities I_p. It can be seen from the figure that the surface quality of the nickel layers at 2 A dm^{-2} and 5 A dm^{-2} current densities are basically similar and the microscopic surfaces exhibit a rhombic vertebral morphology. However, the microscopic surface quality of the nickel layer varied greatly when the nickel sulfate formulations were electrodeposited at different current densities (2 A dm^{-2} and 5 A dm^{-2}), with the microstructure showing columnar grains at $I_p = 2$ A dm^{-2} and finer grains and better surface quality when I_p increased to 5 A dm^{-2}, as shown in Fig. 5.5. Therefore, the process conditions of nickel electrodeposition on the surface of resin model with nickel sulfate formulation and $I_p = 5$ A dm^{-2} were chosen in this book.

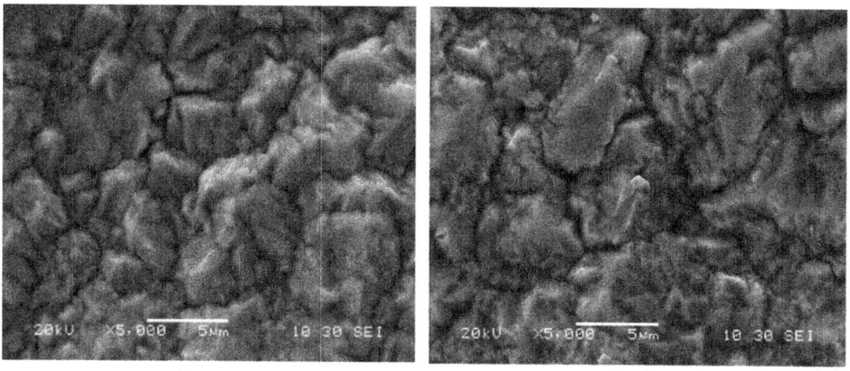

(a) (b)

Fig. 5.4 Electrodeposited microscopic surface quality of nickel sulfamate at different current densities. **a** $I_p = 2$ A dm^{-2}; **b** $I_p = 5$ A dm^{-2}

(a) (b)

Fig. 5.5 Microscopic surface quality of nickel sulfate electrodeposited at different current densities. **a** $I_p = 2$ A dm^{-2}; **b** $I_p = 5$ A dm^{-2}

There are many factors affecting the microscopic surface quality of nickel electrodeposited layers. The composition of deposition solution, current density, additives, DC or pulsed current, PH value, temperature, etc. have effects on the grain size and microstructure of electrodeposited layers, while current density is the most important control factor affecting the surface microstructure during the preparation of crystals by electrodeposition. By changing the current density, not only the speed of electrodeposition but also the cathodic overpotential during electrocrystallization can be changed, thus affecting the nucleation and growth of crystals. It is found that the grain size of electrodeposited nickel layer decreases with increasing current density, and the current density reaches a certain range when it is easy to grow and form nickel nanocrystals.

In addition, many studies have found that the addition of organic additives to the nickel deposition solution can lead to grain refinement. The organic additives added in the electrodeposition process are mainly some surface active substances and brightening agents, such as sodium dodecyl sulfate, saccharin and aliphatic hydrocarbons. The addition of these organic additives has an obvious blocking effect on the electrode reaction, which can increase the cathodic polarization, improve the overpotential of electrocrystallization and effectively refine the grains. The mechanism of action of sugar refined grains [65] is that the saccharin molecules adsorb on the active point of crystal growth during the growth of the deposited layer, which effectively inhibits the crystal growth and promotes the formation of nuclei. Moreover, the hydrogen precipitation on the cathode surface is more obvious at lower pH value, and hydrogen provides more nucleation centers for nickel during the cathodic reduction of nickel ions, which makes the deposited layer of nickel crystallize meticulously and the grains are refined.

5.3 Mechanical Properties Study

5.3.1 Interfacial Bond Strength

Usually, the bonding strength of electrodeposited layer and substrate is measured by bonded tensile method, scratch method, indentation method, wedge loading method and so on. Among them, the bonded tensile method is the most commonly used, and this book adopts the bonded tensile method and measures the bonding strength of resin-nickel layer interface according to ISO-2819 standard. This book uses the bonded tensile method test sample diameter of 20 mm, length of 40 mm, the test sample and the corresponding sample with AB two-component adhesive bonding Instron 1195 universal material testing machine for tensile test, the specimen is shown in Fig. 5.6a is the bonded specimen before tensile, (b) is the fracture surface of the specimen after tensile. Observation of the fracture surface of the test specimen shows that the adhesive bonding agent did not delaminate, which indicates that the adhesive bonding strength is higher than the bonding strength of resin and nickel layer, and the bonding tensile method can effectively measure its bonding strength.

In order to analyze the effect of chemical roughening method and electrodeposition thickness of resin surface on the interfacial bond strength, the test surface of the specimen was roughened according to the chemical roughening process in the previous section, and then tensile testing of the resin-Nickel layer interfacial bond strength was performed. Figure 5.7 shows the stress–strain curve of the bond strength tensile test. As shown in Fig. 5.8, for the same deposited thickness of nickel layer, the difference of interfacial bond strength before and after chemical roughening is obvious, and the interfacial bond strength increases more after roughening. However, for the resin-nickel interfacial bonding interface after chemical roughening, the maximum value of the bonding strength of 11.1 MPa occurred at a nickel layer

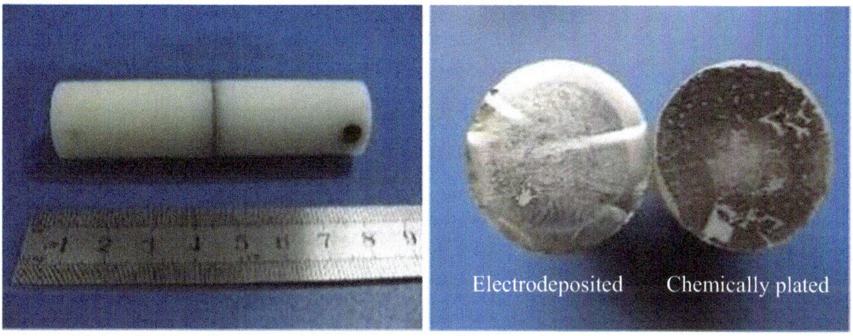

(a) bonded tensile test specimen; (b) typical test specimen fracture surface

Fig. 5.6 Bond strength test specimens and fracture surfaces

thickness of 0.25 mm, while the maximum value of the bonding strength of 5.3 MPa for the resin-nickel interfacial bonding interface without roughening occurred at a nickel layer thickness of 0.15 mm.

This shows that the bond strength does not always increase with increasing thickness of the nickel coating, but decreases with increasing thickness of the deposited layer after reaching a maximum value. This is mainly due to the fact that as the thickness of the deposited layer increases, the residual stresses accumulated on the nickel deposited layer also increase, resulting in a decrease in the bond strength instead.

The surface micromorphology of the resin specimens before and after chemical roughening and nickel chemical deposition and electrodeposition were observed by optical microscopy system (same magnification 1000), as Fig. 5.9 is shown. Among them, (a) and (b) show the surface morphology of the resin model before and after

Fig. 5.7 Bond strength stress–strain curve

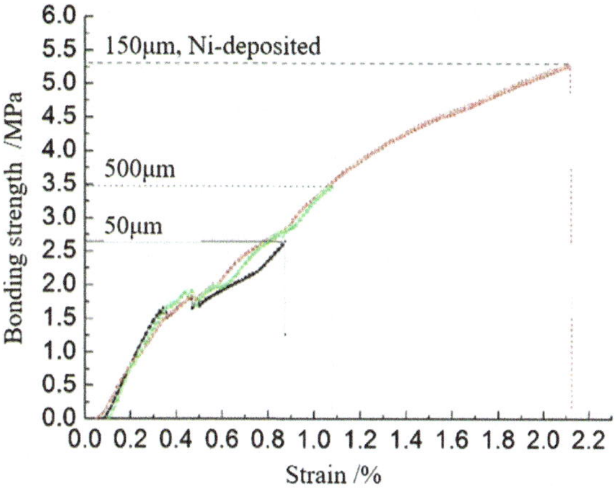

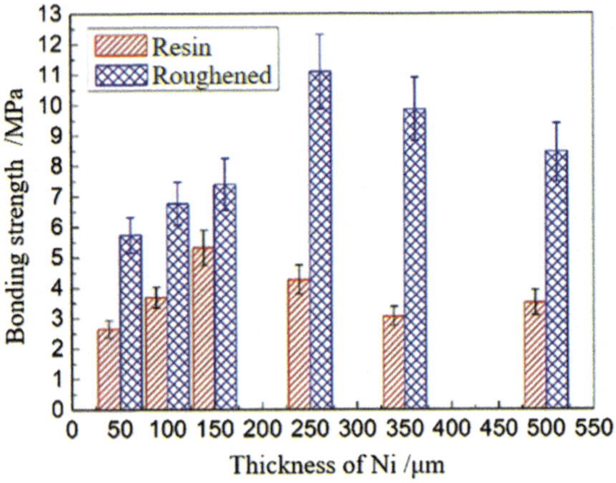

Fig. 5.8 Interfacial bond strength before and after chemical roughening

roughening, and (c) and (d) show the surface micromorphology of nickel chemical deposition and electrodeposition, respectively. From Fig. 5.9 it can be seen that the chemical roughening process on the resin surface is mainly completed by the oxidation of potassium permanganate and chromic acid solution, which results in irregular micro-pores and exhibits the form of porous structure. The experimental results show that some components of the resin material have been eroded by chemical oxidation and formed porous forms in the surface layer. Therefore, the chemically deposited metallic nickel subsequently fills these micro-pore structures until the entire resin surface is formed with a nickel cover layer, as Fig. 5.9c and d show. The location of the mechanical anchor lock between the metal and resin interfaces is mainly present at these microstructure pores, thus increasing the bond strength between the interfaces.

The bonding mechanisms between polymeric material surfaces and metal coatings can be divided into three main categories: mechanical anchor lock, physical bonding, and chemical bonding. Among them, most scholars agree that the mechanical anchor lock effect is the main factor for the increase of bond strength between resin and metal coatings. Physical and chemical bonding effects are difficult to produce between resin and metal coatings, mainly due to the large differences in the physical and chemical properties of the respective component materials themselves, making it difficult to form good physical and chemical bonding effects. Studies have also shown that the surface roughness of the resin matrix material is directly related to the bond strength, which means that the bonding mechanism between the resin-metal interface can be explained by the mechanical anchor-locking effect. Therefore, the micro-pore structure of the resin surface effectively increases the contact surface area between metal and resin, resulting in high and effective bond strength.

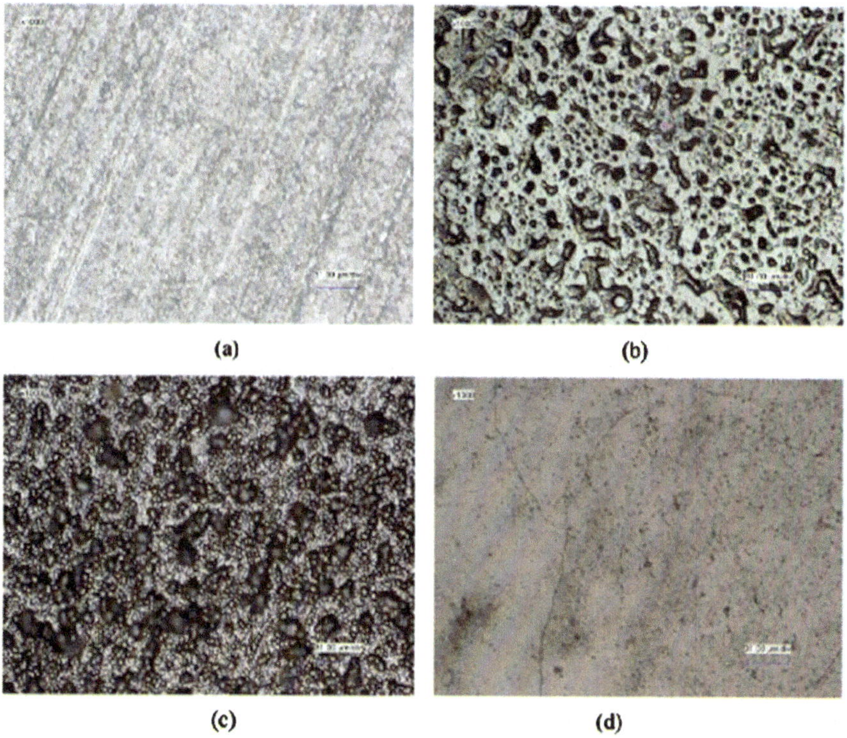

Fig. 5.9 Comparison of the microscopic morphology of the resin surface before and after chemical roughening and its metal deposition effect. **a** Raw resin; **b** resin after chemical roughening; **c** after chemical nickel deposition; **d** after nickel electrodeposition

5.3.2 *Tensile Bending Test*

Since there is no fixed test standard for the mechanical property test of resin-nickel layer composite specimens, this book refers to ISO and ASTM plastic tensile and bending test standards to conduct relevant mechanical tests. According to ISO-527 plastic tensile [47, 48] and ISO-178 plastic bending standards [49]. The tensile and bending specimens of resin-nickel composite are fabricated as shown in Fig. 5.10.

The total length of the light-cured 3D printed resin tensile specimens was 180 mm, and the basic size of the cross-section used for testing was 10 $\times$ 4 mm, and the total length of the bending specimens was 160 mm, and the basic size of the cross-section used for testing was 15 $\times$ 4 mm, and both measured their cross-sectional dimensions accordingly according to the preset thickness of electrodeposited nickel. A universal material testing machine was used for the mechanical property experiments, and the average value was taken from three specimens of each test group.

Fig. 5.10 Mechanical properties of experimental specimens. **a** Tensile; **b** bending

(a)

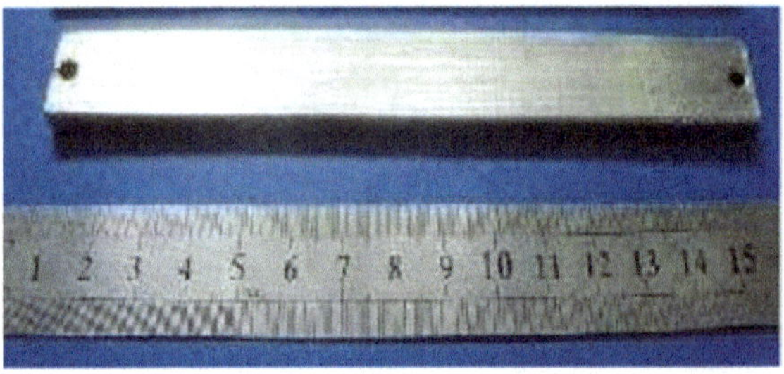

(b)

Figure 5.11 shows the stress–strain curves of test specimens for tensile and bending experiments. Figure (a) shows the comparison of tensile curves of nickel-resin composite specimens with pure resin specimens, and Figure (b) shows the bending stress–strain curves for different thicknesses of nickel electrodeposition. From Fig. 5.11a it can be observed that the stress value of the electrodeposited nickel-resin composite specimen is larger than that of the pure resin at the same strain (2.5%), indicating that it has a higher strength to carry tensile strength, but it fractures immediately after exceeding this strain, showing brittle material properties. From Fig. 5.11b a jump in stress can be observed in the middle section of the bending test.

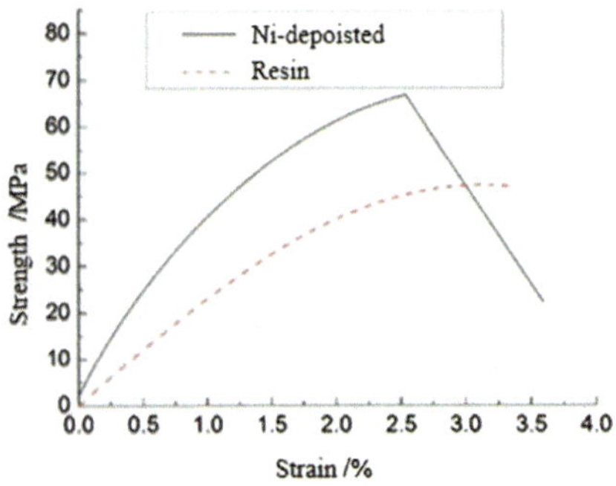

(a)

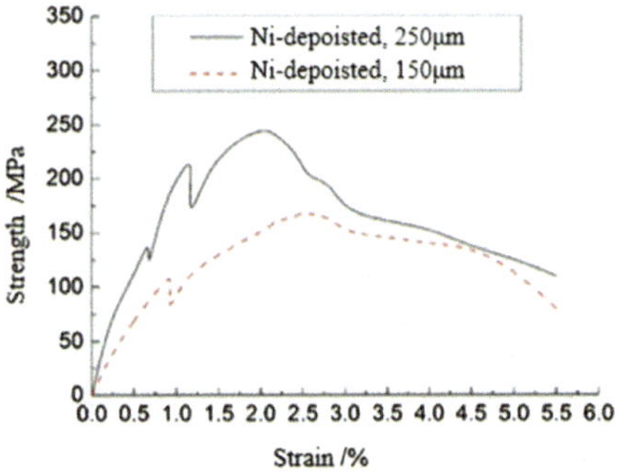

(b)

Fig. 5.11 Stress–strain curve of the mechanical performance experiment. **a** Tensile test; **b** bending test

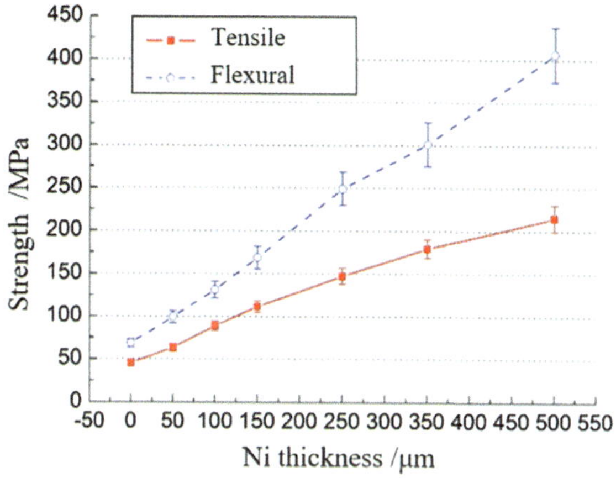

Fig. 5.12 Tensile and flexural strength of composite specimens

The tensile and bending strength curves of the nickel-resin composite specimens show that both tensile and bending strengths increase continuously with the increase of deposition thickness as shown in Fig. 5.12. The tensile and bending strengths of the composite specimens were 89 MPa and 131.6 MPa, respectively, when the thickness of nickel deposited was 0.1 mm, which were 1.95 times and 1.91 times higher than the tensile strength (45.7 MPa) and bending strength (68.9 MPa) of the pure resin material, respectively. When the thickness of nickel layer was 0.5 mm, the tensile and flexural strengths of the composite specimens were increased by 4.7 and 5.9 times, respectively, compared with that of the pure resin material.

The curves of the tensile and flexural elastic models of the nickel-resin composite specimens with the deposition thickness are shown in Fig. 5.13, and it can be seen from the figure that the elastic modulus increases with the increase of the deposition thickness. When the thickness of nickel deposition is 0.1 mm, the Young's modulus and bending modulus of the composite specimen are 10 GPa and 15.4 GPa respectively, which are 4 times and 6.7 times higher than the Young's modulus (2.5 GPa) and bending strength (2.3 GPa) of the pure resin material. When the thickness of nickel layer is 0.5 mm the tensile and flexural modulus of composite specimens are increased by 14.5 and 22.3 times, respectively, compared with the pure resin material.

Figure 5.14 shows the macroscopic fracture morphology of the resin-nickel layer. Figure 5.15 shows the microscopic fracture pattern of the electrodeposited nickel layer. Looking at the fracture of the tensile specimen, it can be seen that the electrodeposited nickel layer exhibits a typical ductile fracture. The fracture of the nickel layer under tensile load is macroscopically rough, grayish, fibrous, with a 45° shear lip at the edge of the part surface, and its microscopic characteristics are mainly the fracture surface consists of some small pits, which are actually grown-up hollow nuclei, i.e. tough nests. Tough nest fracture [78] is a ductile fracture belonging to a high-energy absorption process, which proceeds through the formation of cavity nuclei grown and interconnected. Moreover, the shape of the tough nests after fracture

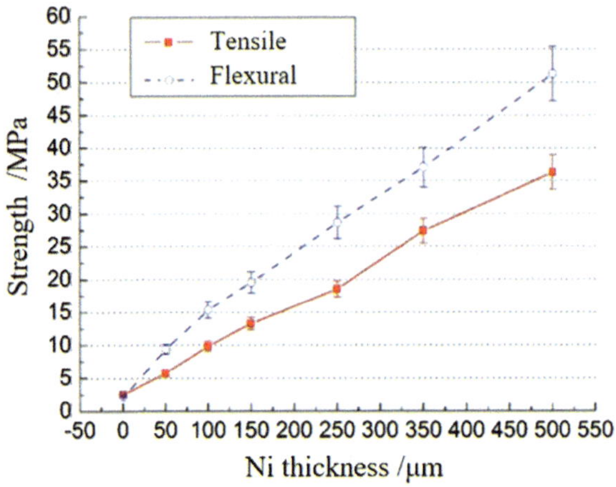

Fig. 5.13 Tensile and flexural modulus of elasticity of composite specimens

of the electrodeposited nickel layer exhibits round and elliptical shapes of inconsistent size, which grow uniformly around the microscopic cavities under the tensile positive stress and form nearly circular isometric tough nests after fracture. Since the cured Somos 14120 resin material itself is a toughened brittle material, it mainly exhibits brittle fracture during the tensile process. The fracture surface of its brittle fracture is close to perpendicular to the tensile stress, and the macroscopic fracture mainly consists of crystalline bright surfaces with a glossy finish.

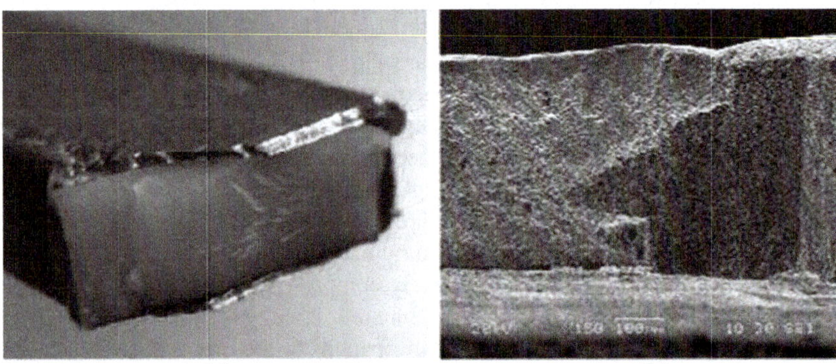

(a) (b)

Fig. 5.14 Macroscopic fracture morphology of resin-nickel layer. **a** Resin-nickel layer macroscopic fracture; **b** electrodeposited nickel layer fracture

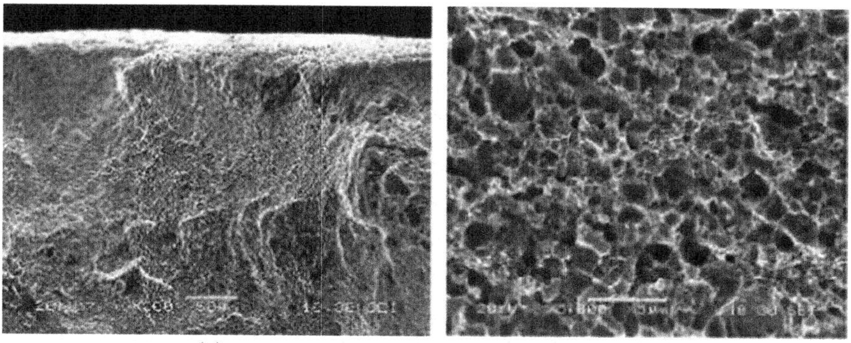

(a) (b)

Fig. 5.15 Microscopic fracture morphology of electrodeposited nickel layer. **a** Fibrous fracture morphology; **b** tough fossa

5.3.3 Analysis and Discussion

1) Effect of interfacial bond strength on mechanical properties

From the fracture surface of the typical bond strength test specimen in the previous section, it can be seen that the interfacial bond layer between the resin matrix and the chemically deposited nickel layer does not fall off extensively, but is more tightly bonded to the resin matrix, which can indicate that the roughening effect of the resin surface is more obvious. The fracture surface produced by the bonding and tensile method is mainly at the interface between the chemically deposited nickel layer and the electrodeposited nickel layer, and the bond strength stress–strain curve also shows that the interface fractures very rapidly. Therefore, the interface between the electrodeposited resin and the nickel layer can be divided into two layers: (a) the interface between the resin matrix and the chemically deposited nickel layer; (b) the interface between the chemically deposited nickel layer and the electrodeposited nickel layer. The interfacial bonding strength between the chemically deposited nickel layer and the electrodeposited nickel layer is the main factor affecting the mechanical properties. The effect of resin-metal interfacial bond strength on mechanical properties is mainly manifested in tensile Young's modulus and bending strength.

(1) Young's modulus

According to the theoretical calculation method of Young's modulus of composite materials, we can get the calculation formula of resin-nickel layer composite specimen [50]:

$$E_c = E_m V_m + E_s(1 - V_m) \tag{5.1}$$

Fig. 5.16 Comparison of theoretically calculated and experimentally measured modulus of elasticity

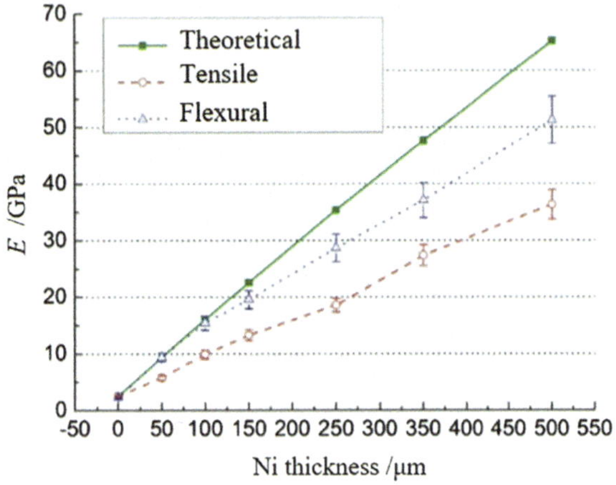

Eq

E_c Young's modulus of the composite specimen/GPa;
E_m Young's modulus of electrodeposited nickel/GPa;
V_m volume percentage of nickel layer/%;
E_s Young's modulus of resin matrix/GPa.

The Young's modulus of the composite specimen was calculated and compared with its experimental modulus by Fig. 5.16a, the Young's modulus of the resin-nickel composite specimen measured experimentally is much smaller than the theoretical calculated value, mainly because the actual resin-nickel layer interface bond strength is lower, while the premise of the theoretical calculation assumes that the resin-metal interface is completely tightly bonded, and the interface is not affected by tensile stress during the tensile process, thus causing the theoretical value of Young's modulus of the composite specimen to be larger than the experimental value.

(2) Bending performance

During the bending experiments of composite specimens, the bending stress–strain curve appears to jump in the middle section as a typical case. The main reason for this may be due to the lower interfacial bond strength of the bending specimen and the form of stress jump curve. For the resin-nickel composite specimen, during the bending process, the form of force on the specimen is more complex, showing the combined effect of compressive stress, shear stress and torsional stress, and the interface bonding part is affected by various forms of external forces. The bending stress jumps in the middle section may be due to the resin-nickel interface in a variety of complex forms of external forces began to detach or fracture, but the external nickel layer can still resist the action of bending stress, composite specimen bending stress increases with strain, until the nickel layer fracture.

Moreover, the bending performance of the composite specimens increased more significantly than the tensile performance, which is mainly due to the fact that the bending test has a greater effect on the interfacial bond strength than the tensile test and is related to the cross-sectional area of the specimens. Therefore, the interfacial bond strength has a certain effect on both tensile and bending properties of composite specimens, and the effect on bending properties is particularly significant.

2) Effect of nickel volume percentage on mechanical properties

The volume percentage of the nickel layer V_m is calculated using the formula

$$V_m = \frac{2d \times (w + h + 2d)}{(h + d) \times (w + d)} \times 100\% \tag{5.2}$$

Eq

d Nickel layer thickness/mm;
w Width of resin specimen/mm;
h height of resin specimen/mm;
V_m Nickel layer volume percentage/%.

If $w + d = p$ and $h + d = q$, the percentage volume of nickel layer V_m:

$$V_m = \frac{2d \times (p + q)}{p \times q} \times 100\%. \tag{5.3}$$

From Eqs. 5.2 and 5.3, it can be seen that for the same thickness of nickel electrodeposited layer, the percentage of nickel layer decreases with the increase of specimen cross-section. The volume percentage of nickel layer is not only related to the thickness of nickel layer, but also related to the size of specimen cross section. The mechanical properties of the composite specimen mainly depend on the strength and stiffness of the nickel layer, so when the thickness of nickel layer is certain, the strength and stiffness of the composite specimen mainly depend on the volume percentage of the nickel layer, the larger the resin substrate cross-section is, the smaller the volume percentage of its nickel layer, the smaller the strength and stiffness of its composite specimen. Therefore, the effect of electrodeposition to improve the mechanical properties of the resin wind tunnel test model depends on the thickness of the nickel layer and the cross-sectional size of the model part.

Since the accuracy of electrodeposition decreases with the increase of thickness, in order to ensure the accuracy of wind tunnel test model, the thickness of nickel layer of resin wind tunnel test model for electrodeposition is generally taken as 0.1–0.2 mm. When the thickness of nickel layer is certain, the strength and stiffness of resin-nickel layer composite wind tunnel test model mainly depends on the size of model cross-section. Therefore, when electrodeposited nickel is used to improve the strength and stiffness of the resin wind tunnel test model, the electrodeposition method is only applicable to thin wind tunnel test model parts such as wing, aileron, flap, tail, etc. with small cross-sectional size.

Moreover, for a given cross-section of the matrix, the mass and stiffness of the resin-nickel layer composite specimen increase with the increase of the nickel deposited layer. However, if the cross-sectional size and form of the resin matrix are changed, the mass and stiffness of the composite are related to the volume percentage of each component material, i.e., the material volume composition of the resin and nickel layers is related to the mass and stiffness. Therefore, in order to obtain a suitable mass distribution and suitable specific stiffness, it is crucial to design the thickness and cross-sectional proportional relationship between resin and nickel layers, which is relevant for the design and fabrication of dynamically similar wind tunnel test models (static aeroelastic, fluttering models) requiring similar mass distribution and stiffness.

5.4 Electrodeposition Wing Pressure Measurement Model Manufacturing

The manufacturing process of composite wing pressure measurement model assembly by electrodeposition can be divided into the following steps: (1) modify or add wire process holes according to the electrodeposition process requirements; (2) compensate the CAD model for inward offset according to the preset electrodeposited nickel layer thickness; (3) manufacture the resin model after inward offset using 3D printing equipment; (4) rough the model surface; (5) conductive the model surface treatment; (6) electrodeposition of high-strength nickel metal of preset thickness.

As the electrodeposited nickel layer has a certain thickness, after the parts are surface treated, it will inevitably cause changes in the dimensions of the parts. Usually the model dimensions and tolerances specified on the design drawings refer to the final dimensions and tolerances of the model before electrodeposition, and the thickness of the deposited layer and its electrodeposition dimensional deviation should be reserved in advance in conjunction with the final dimensions. The assembly surface not only has the offset and the deposition thickness as twice, but also has the influence of the assembly error, and the model offset should be considered as a whole when designing. Since the thickness of chemical deposition is generally negligible (1–5 μm), the inward offset compensation of 3D printing-based electrodeposition is about 0.1–0.2 mm. Chemical deposition has basically 100% dispersion capability because there is no current distribution on the part, and a uniform thickness of deposited layer can be obtained regardless of deep holes, blind holes, deep grooves or complex shaped workpieces. Figure 5.17 shows the chemical deposition effect of the wing pressure model and the local enlargement of a pressure hole.

The electrodeposition of the nickel layer on the model surface exhibits convexity and unevenness due to the effect of the electrolyte dispersion ability under the action of external electric field. The geometric factors are the shape of the electrodeposition bath, the shape of the anode, the shape of the part and the mutual position and distance between the part and the anode, etc. The geometric factors lead to tip discharge and

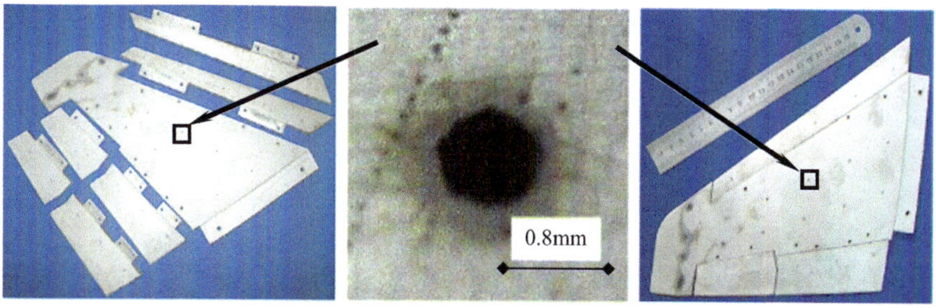

Fig. 5.17 Chemical deposition effect of wing pressure measurement model

edge effect, which affect the current distribution in the electrodeposition process. Electrochemical factors include polarization, current density, conductivity of the solution, and current efficiency. Therefore, the following methods are mainly adopted to improve the uniformity of the nickel layer during the electrodeposition of the wing pressure measurement model: (1) reasonable mounting of the parts, with the main surface of the model facing the anode and parallel to it, so that the parts are under the best current distribution; (2) adjusting the distance between the cathode and the anode according to the possibility, within a certain range, increasing the distance between the cathode and the anode to improve the dispersion ability, reducing the distance ratio between different parts of the cathode and the anode distance ratio between cathodes and anodes; (3) using double anodes to improve the current distribution; (4) using cathode shielding to reduce the current density in the concentrated part of the power line. Figure 5.18a shows the results of the edge effect of electric field distribution due to the formation of the leading edge corners of the wing, Fig. 5.18b shows the effect of electrodeposition of the wing model after improving the power line distribution by using parallel mounting and double anodes. Figure 5.19 shows the before and after electrodeposition effect of the scaled-down model of the whole aircraft.

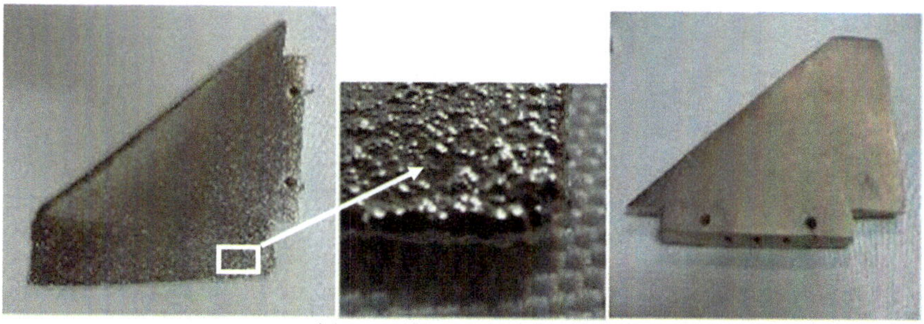

(a) (b)

Fig. 5.18 Comparison of the effect of electrodeposition wing model before and after electric field improvement. **a** Edge effect of electric field distribution in the wing model; **b** electrodeposition effect after electric field improvement

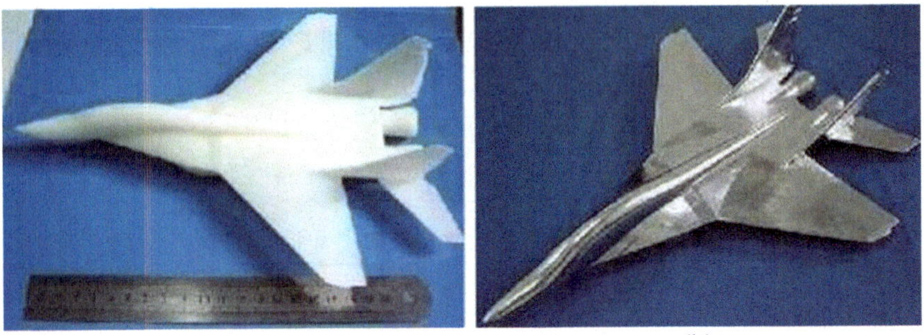

(a) (b)

Fig. 5.19 Electrodeposition of the whole machine model to enhance the manufacturing effect. **a** Resin model; **b** Electrodeposition model

An example of the fabrication of a wing pressure measurement model with nickel electrodeposition, after testing the pressure measurement orifice is still conductive after electrodeposition, as shown in Fig. 5.20a shows each of the electrodeposited parts that make up the wing model. Figure 5.20b is the pressure measurement hole on the main wing tip surface. Each front flap, aileron and variator are printed in one piece with 0° and ± 20° deflection angles, and the mounting part of the variator is embedded inside the wing without destroying the wing shape and assembled by pin positioning. As the cross section of the assembly part of the model forms multiple layers of electrodeposited nickel after electrodeposition, it is very helpful to increase the strength of the assembly connection. Figure 5.20c shows the assembly effect of the electrodeposited wing pressure measurement model.

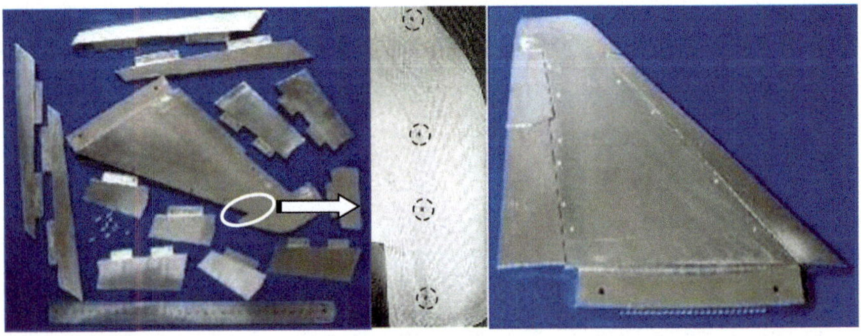

(a) (b) (c)

Fig. 5.20 Example of fabrication of wing pressure measurement model with nickel electrodeposition. **a** Wing pressure model assembly; **b** surface pressure holes; **c** wing pressure model assembly effect

5.5 Model Manufacturing Economics Analysis

The manufacturing of metal wind tunnel test models by traditional machining methods is a complex process, in which the process design process takes a lot of time, and the process design time of a domestic set of models accounts for about 25–30% of the whole production period. Table 5.1 is the typical process flow of wing force measurement model. The process is characterized by many processes and low efficiency. If we use light-curing 3D printing technology, we only need to make certain corrections to the CAD model structure design and data processing to enter the printing manufacturing. Light-curing 3D printing makes it very convenient to make recesses, convex shoulders and hollow parts of the model without special tooling and tool design, and without the preparation of numerous processes and processing materials, saving a lot of labor, material and time costs. Moreover, when processing slender models such as wings, there is no need to worry about machine vibration and cutting forces that may cause deformation.

The wing force measurement model (320 × 350 × 30 mm) for engineering applications and the pressure measurement (210 × 240 × 16 mm) test model manufactured in this book were used for manufacturing economy analysis. The volume of the force measurement and pressure measurement models were approximately 4 times related, and their models were uniformly scaled to the force measurement model for time and cost comparison. The light-curing printing time can be predicted using parameters such as model volume, effective surface area, and support structure factors. The printing time for the force model is about 16–18 h, and the printing time for the pressure model is about 10–12 h. Under the same scaling ratio and the same printing conditions, the manufacturing time should be similar, so it is taken as 16–18 h. The post-processing time for the wing force measurement model, such as cleaning, surface grinding and post-curing, is about 1–2 h. As there are pressure measurement holes in the force measurement model for cleaning, and the more holes arranged the more difficult it is to handle, and the more detailed the surface finishing, the post-processing time will be more, about 3–5 h. The time spent for chemical deposition is about 2 h, and the electrodeposition time is related to the current density and the thickness of the deposited layer. If the current density is 1 A/dm^2, the time required for electrodeposition of nickel layer thickness of about 150 μm is 13 h.

The time comparison of each manufacturing method is analyzed as Table 5.2. Compared with CNC machining, the time for manufacturing wing force model is reduced by about 60% for 3D printing technology and 80% for wing pressure model. For the numerical control machining of the pressure measurement model, the smaller the model, the more difficult it is to manufacture, especially the pressure measurement hole machining of the flaps and so on, the longer the manufacturing time and the higher the manufacturing cost, so there is a great advantage of using 3D printing

Table 5.1 Typical machining process documents for wing force measurement model

Serial number	Type of work	Process content and requirements
1	Tongs	Laying out holes and deburring according to the sketch outline size (2 pieces)
2	Planning	Plan two major surfaces to see the light parallel can be, plan and its adjacent two right-angle surface P, Q surface perpendicular to each other (perpendicularity 0.05)
3	Grind	Grinding two major surfaces parallel to the light
4	Cut	Cut the wing shape according to the drawing (including both end process blocks)
5	Milling	Precision milling of the wing profile according to the numerical model, leaving a margin of 0.1 mm outside the profile
6	Inspection	Coordinate inspection of profile data and mating dimension tolerances
7	Tongs	Draw the wing chord plane line, sample stand line, fine down the sample according to the wing sample, fine repair the external surface And streamline sanding, according to the map of the front edge of the sagging groove line and each variable angle piece groove line
8	Milling	Precision milling of sunken grooves and each corner piece groove according to the drawing and reference scribe lines
9	Tongs	Repair 0° angle piece, transition between angle piece appearance surface and wing streamline, match each hole, draw each flap and aileron line
10	Cut	Cut the leading and trailing edge flaps according to the diagram and line (to ensure the size of the main body of the wing)
11	Milling	Precision milling of trailing edge flap ailerons according to drawing and digital model
12	Cut	Cut off process blocks according to the diagram and lines
13	Tongs	Assembling with the body, repairing the connection block, mounting groove and type surface, and making each connection hole

manufacturing method. The above manufacturing time and cost analysis is only at the experimental research stage. If the rapid manufacturing method of wind tunnel test model is improved into a unified manufacturing specification and the manufacturing links are specialized, the rapidity and accuracy of wind tunnel test model manufacturing can be systematically ensured.

Table 5.2 Wind tunnel test model manufacturing time analysis table

Model type	Manufacturing method	Time/h
Wing force measurement model	Light-curing SL	16–18/1–2 (print/post-processing)
	Electrodeposition ED	2/13 (chemical deposition/electrodeposition)
	CNC machining NC	40–50
	SL versus NC	50–65% reduction
	SL + ED relative NC	15–35% reduction
Wing pressure measurement model	Light-curing SL	16–18/3–5 (print/post-processing)
	Electrodeposition ED	2/13 (chemical deposition/electrodeposition)
	CNC machining NC	100–120
	SL versus NC	75–85% reduction
	SL + ED relative NC	60–70% reduction

Chapter 6
Additive Manufacturing of Wind Tunnel Test Models for Force Measurement

6.1 Introduction

The force measurement model is the basic model in the wind tunnel test, used to test the performance of the aerodynamic shape, the geometric similarity of the shape is important to the test quality, and the safety and processing efficiency of the model has an important impact on the wind tunnel test scheduling. The design and processing of the wind tunnel test model is an important part of the aircraft wind tunnel test, which has an important impact on the development cycle and cost of the aircraft [18]. The design and processing of wind tunnel test models is an important part of aircraft wind tunnel testing, and has a significant impact on the development cycle and cost of aircraft. At present, the practical models are realized by metal mechanical processing. Metal (high quality steel or hard aluminum) has excellent and stable mechanical properties, which can ensure the safety and stability of the wind tunnel test; the manufacturing technology represented by multi-degree of freedom CNC machining can provide satisfactory machining accuracy. Therefore, the metal-machining mode is used for the design and processing of high-performance models [3, 4]. However, the advancement of advanced manufacturing technology and material science provides the possibility to rethink the traditional model and develop new model design and processing methods. On the one hand, wind tunnel test models are a special class of industrial products serving aircraft development, with typical single-piece and small batch characteristics. Aircraft development will propose a variety of options for aerodynamic layout, shape parameters, weapon configuration, internal structure, etc., and require a large number of wind tunnel tests, therefore, the design and manufacturing of models should be able to quickly respond to the needs of aircraft designers; however, this is not the case for On the other hand, non-metallic materials (polymers, composites, etc.) have made great progress and have been able

W. Zhu and D. Li, *Models for Wind Tunnel Tests Based on Additive Manufacturing Technology*, Additive Manufacturing Technology,
https://doi.org/10.1007/978-981-99-5877-1_6

to provide a wider choice of materials for engineering applications. Compared with metals, polymers and composites usually have better specific strength and specific stiffness, although they have lower strength, which is of great value for designing and processing lightweight wind tunnel test models [43, 44]. Meanwhile, the lower strength and stiffness also provide new possibilities for the development of aircraft models, such as Yan Yin [66, 67], Qianwe [15, 66, 67] Yang Zhichun [13] and Li Duchen [54] and other groups have verified that the low modulus of the material helps to achieve the design and fabrication of structurally similar elastic models, which is not possible using metals as the model material.

In recent years, several departments at home and abroad have conducted research on 3D printing technology to innovate the design and manufacture of wind tunnel test models for aircraft. Attention was first focused on the feasibility study of replacing metal models with non-metal models. NASA's Springer team [17, 25] NASA's Springer team, Alabama University's Landrum team [39]. The team of Springer at NASA, the team of Landrum at Alabama University, USA, the team of Chuk at McGill University, Canada [24]. The first feasibility studies were carried out in the 1990s by Springer's team at NASA, Landrum's team at Alabama University, Chuk's team at McGill University, Canada, and Azarov's team at the Central Pneumatic Research Institute (TsAGI), Russi [68]. The first studies were conducted in the 1990s and demonstrated the feasibility of 3D printing technology for the design and fabrication of wind tunnel test models in preliminary aerodynamic studies. The USAF has used the method for the E-8C early warning aircraft [30] X-45A UAV [34, 69] and so on, and achieved good results. However, most of the above studies have used simple aircraft structures as objects [17, 25, 44, 70] (missiles, aircraft wings, standard models, etc.), and rarely involve models with complex detailed features (e.g., hanged bombs, adjustable rudder surfaces, etc.). The design and machining capabilities of these features are important for practical models and pose further challenges to the feasibility of 3D printing technology.

Metal force measurement models based on traditional machining methods have difficulties in detail simulation, many parts, heavy models, and long processing cycles. This chapter develops a new method for the design and fabrication of high-similarity force measurement models based on additive manufacturing technology, taking advantage of its seamless virtual data-physical part conversion, the integrated printing capability of complex structures, and the low-density characteristics of plastic materials. In this chapter, the plastic-metal hybrid structure scheme combines the advantages of additive manufacturing technology for rapid processing of complex aerodynamic shapes and internal mounting structures and the characteristics of machined high-precision parts to ensure higher geometric similarity. The overall technical scheme is shown in Fig. 6.1.

With the powerful internal and external structure printing capability of the light-curing printing technology, the model can be designed to be lightweight in order to improve the vibration safety of the model in the wind tunnel. The model includes aircraft prototype detail features such as hang-ups, fuselage bulges, adjustable rudder surfaces, etc. Due to the introduction of light-curing printing technology, the structure of the model has been simplified, the number of parts has been reduced, and the

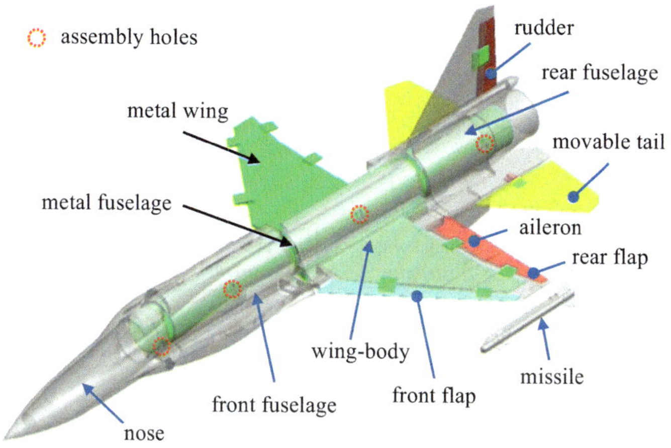

Fig. 6.1 Structure of resin-metal composite wind tunnel test model

geometric similarity of the model has been improved. Based on the characteristics of the light-curing printing technology, this book designs the available rudder surface deflection mechanism in detail for different rudder surface structures, and proposes a resin hole reinforcement scheme to ensure the accuracy of resin-metal assembly. The strength calibration of the model is carried out by numerical calculation, and finally the model is tested in wind tunnel and compared with the data of the metal model.

6.2 Lightweight Design

6.2.1 Optimization Strategies

As mentioned above, the aim of this chapter is to design and fabricate a lightweight rigid model, where weight reduction design is a central issue. The excellent printing capability of SL technology and the low density of the materials it uses [24, 71]. The excellent printing capability of SL technology and the low density of the materials used offer new possibilities for weight reduction design. On the one hand, the weight of the model can be reduced to a great extent by replacing the heavier metal with lightweight resin to make the shell of the model; on the other hand, the structure of the metal parts can be minimized and their machinability improved, as they only play a supporting role inside the model and are not involved in the composition of the aerodynamic shape, which helps to improve the cycle time and cost of the model design and processing.

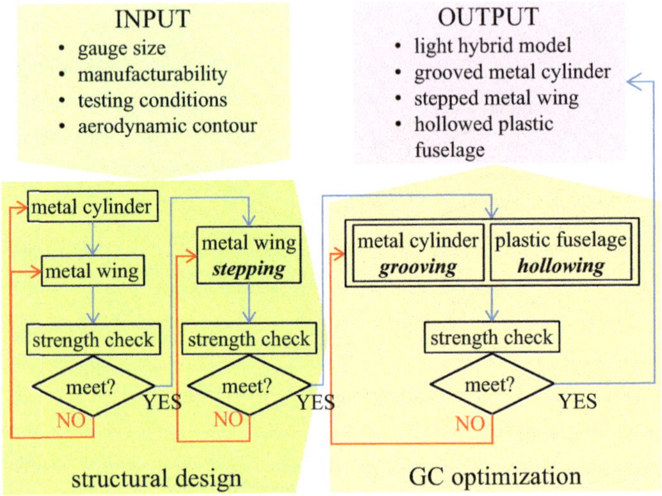

Fig. 6.2 Design flow chart of the lightweighting composite model

As shown in Fig. 6.2, the entire design process can be divided into two major steps: preliminary design and structural optimization. The initial structure of the resin shell is obtained by subtracting the metal skeleton from the aerodynamic profile geometry; weight reduction In this step, the structures of the metal skeleton and the resin shell are further optimized to reduce the total weight of the model.

6.2.2 *Preliminary Design*

The basic structure of the metal skeleton and resin shell will be determined in this step. The basic dimensions of the body column include the inner wall diameter, wall thickness and column length, and the lightweight model requires these three dimensions to be as small as possible. The inner wall diameter of the column is limited by the test balance, which requires a solid assembly of the column and the balance on the one hand, and the book is designed with a conical connection structure as shown in Fig. 6.3. The length of the column should be as short as possible while meeting the installation requirements of the wing skeleton, resin housing, etc. For the wing skeleton, the flat shape size and thickness are determined by the attachment structure—resin shell, the installation of control surface and the loading requirements of the model. The initial structure of the resin shell is obtained on the basis of the above-mentioned skeleton structure, and the excellent capability of SL technology to print complex structures makes the design of the resin shell relatively simple, simply by Boolean subtraction of the model's form geometry from the skeleton geometry.

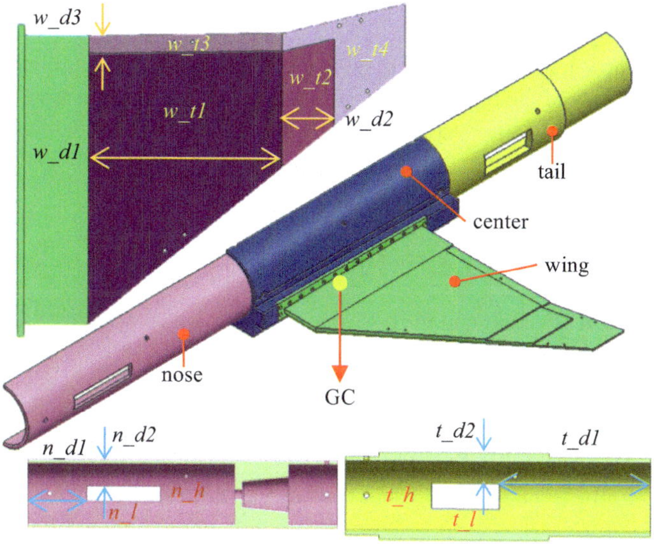

Fig. 6.3 Structure of the metal skeleton and its optimal configuration

The obtained initial model is statically analyzed using Finite Element Analysis (FEA) to obtain its strength and stiffness data. With the model's sufficient load carrying capacity as the constraint variable, the initial model (e.g., a flat plate of the wing) can be optimally designed to further reduce the weight of the model. As shown in Fig. 6.3, the flat plate structure of the metal wing is further optimized into a step structure, which is more efficient and naturally lower in weight compared to a flat plate of uniform thickness due to the loading characteristics of the cantilever beam of the wing. For this reason, the corresponding structural optimization mathematical model needs to be established and the optimal solution is obtained using the automatic optimization method. This structural optimization problem can be expressed as the following mathematical model:

$$\begin{cases} \text{minimize} & f(\boldsymbol{x}) \\ \text{subject to} & \boldsymbol{x}^l \leq x \leq \boldsymbol{x}^u \end{cases} \tag{6.1}$$

1) Objective function

The goal of optimization is to obtain the minimum weight of the structure (here, a metal wing flat plate is used as an example), so the objective function can be expressed as the following function:

$$f(\boldsymbol{x}) = w_{\text{MetalWing}}(\boldsymbol{x}) \tag{6.2}$$

2) Optimization variables

As Fig. 6.3 shown, the metal wing plate is divided into several optimization regions with the thickness of the regions increasing along the trailing edge-tip to the leading edge-root. The optimization variables include the thickness of the region and the positioning of the region. The optimization variables can be expressed as the following vectors:

$$\boldsymbol{x} = (\boldsymbol{x}_{\text{Thickness}}, \boldsymbol{x}_{\text{Distance}}) \tag{6.3}$$

where the thickness vector and the positioning vector are composed of the following dimensional scalars:

$$\boldsymbol{x}_{\text{Thickness}} = (w\_t1, w\_t2, w\_t3, w\_t4) \tag{6.4}$$

$$\boldsymbol{x}_{\text{Distance}} = (w\_d1, w\_d2, w\_d3, w\_d4) \tag{6.5}$$

3) Binding variables

The purpose of setting the constraint variables is to ensure that the optimization results are harmonious and process-oriented. For example, load-bearing requirements, minimum and maximum size requirements, etc. In this chapter, the worst test conditions are used as the load accounting conditions; the minimum size is set based on processability, generally greater than 0.5 mm; the maximum size of the embedded metal skeleton does not exceed the shell shall prevail.

6.2.3 Structural Optimization

Structural optimization is based on the preliminary design, the structure of the model is further optimized to further reduce the weight of the model while ensuring that the center of gravity of the model is in the design position. The main objects of optimization in this step are the metal fuselage column and the resin shell at the fuselage, where the excess weight of the former is removed by means of grooving, and the unnecessary internal entities of the latter are hollowed out, as shown in Fig. 6.3. The capacity of 3D printing technology to print complex internal structures increases the freedom of weight reduction design and is the basis for lightweight design of the model. The optimization calculations in this section are modeled as the following mathematical expressions:

$$\begin{cases} \text{minimize} & f(\boldsymbol{x}) \\ \text{subject to} & \boldsymbol{x}^l \leq x \leq \boldsymbol{x}^u \end{cases} \tag{6.6}$$

1) Objective function

The minimization of the total weight of the model is the aim of the design of this section, so the objective function can be expressed as.

$$f(\boldsymbol{x}) = w_{\text{Model}}(\boldsymbol{x}) \tag{6.7}$$

2) Optimization variables

As Fig. 6.3 shown, the optimization variables are the parameterized dimensions of the participating design components (e.g., metal head, metal tail, and resin head). Thus, the total optimization variables vector can be expressed as

$$\boldsymbol{x} = (\boldsymbol{x}_{\text{MetalNose}}, \boldsymbol{x}_{\text{MetalTail}}, \boldsymbol{x}_{\text{PlasticNose}}) \tag{6.8}$$

This project reduces the weight of the metal body by cutting slots in it. The parameterized dimensions include the location and size of the slots, so the variable vectors for the metal head and tail are as follows:

$$\boldsymbol{x}_{\text{MetalNose}} = (n\_l, n\_h, n\_d1, n\_d2) \tag{6.9}$$

$$\boldsymbol{x}_{\text{MetalTail}} = (t\_l, t\_h, t\_d1, t\_d2) \tag{6.10}$$

Figure 6.4 shows the scheme diagram of the hollowed-out design of the resin shell. The wing and the adjacent fuselage section were not involved in this round of structural optimization for safety reasons due to the large load capacity near the wing. Other than that, all resin shell parts were hollowed out, including the resin nose, resin center fuselage and resin tail. Among them, the center fuselage and tail retain the safety wall thickness of 2 mm ~ 5 mm and undergo a non-parametric shell extraction process, which is no longer involved in the subsequent optimization calculation; the hollowing of the resin nose includes both parametric and non-parametric parts, and the parametric hollowing is involved in the optimization calculation. As shown in Fig. 6.4, the faces marked in green are obtained by the non-parametric shell extraction operation, while the blue faces are parametrically hollowed. The parametric hollowing operation of the head is controlled by three circular sections: front, middle and rear, where the largest rear section has fixed dimensions due to its fit with the metal head, and the dimensions and positions of the other two circular sections can be used as optimization variables with vector expressions as

$$\boldsymbol{x}_{\text{PlasticNose}} = (pn\_r1, pn\_r2, pn\_d1, pn\_d2) \tag{6.11}$$

3) Binding variables

Again, a reasonable range of dimensions is the purpose of setting the constraint variables, and the principles for setting them are the same as in the previous section.

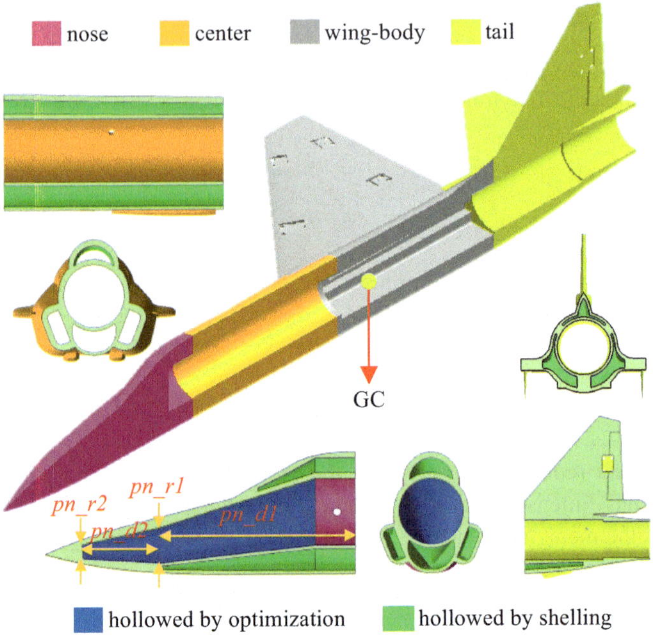

Fig. 6.4 Hollowing of the resin shell and its optimal configuration

In addition, a constraint on the position of the total center of gravity of the model is added to this round of optimization.

6.3 Adjustable Rudder Surfaces

The rudder surface mechanism includes the flaps and ailerons of the main wing, the rudder and elevator of the drogue and flat tail, and other operating surfaces, which are the core to ensure the aerodynamic performance and realize the flight control of the aircraft. In the development of aircraft, there are a large number of models need to be equipped with adjustable deflection angle rudder mechanism.

6.3.1 Steering Surface Adjustment Mechanism

The rudder surface manipulation mechanism to change the rudder surface declination angle, both convenient and accurate, and to ensure that during the test, each manipulation surface declination angle does not change due to aerodynamic action [4]. Therefore, this book has designed a new design for different rudder surfaces.

Therefore, this book designed the corresponding rudder surface adjustment mechanism for different rudder surfaces, divided into the rudder surface of the main wing and vertical tail and full-motion flat tail.

1) Main wing and vertical tail

The installation position of the main wing and drogue is larger, and the adjustment method of variable angle piece can be used. As Fig. 6.5c shows, a series of angle pieces are machined according to the test content, and the change of rudder declination is realized by changing different angle pieces. In this way, the surface of the model is clean, the connection is reliable, and the angular repeatability is high [4]. The model has clean surface, reliable connection, and high repeatability of angle positioning.

Figure 6.5a shows the rudder control mechanism on the main wing. The metal variole is screwed to the metal skeleton inside the resin main wing, while the rudder surfaces (ailerons, front and rear flaps) are screwed to the variole. In the experiment, different deflection angles of the rudder surface can be achieved by simply changing the vario-angle piece. The drogue tail (Fig. 6.5b) also adopts a similar design scheme, except that there is no metal skeleton inside the resin drogue, so the variator is directly installed in the resin mounting holes. Because of SL's powerful printing capability, all

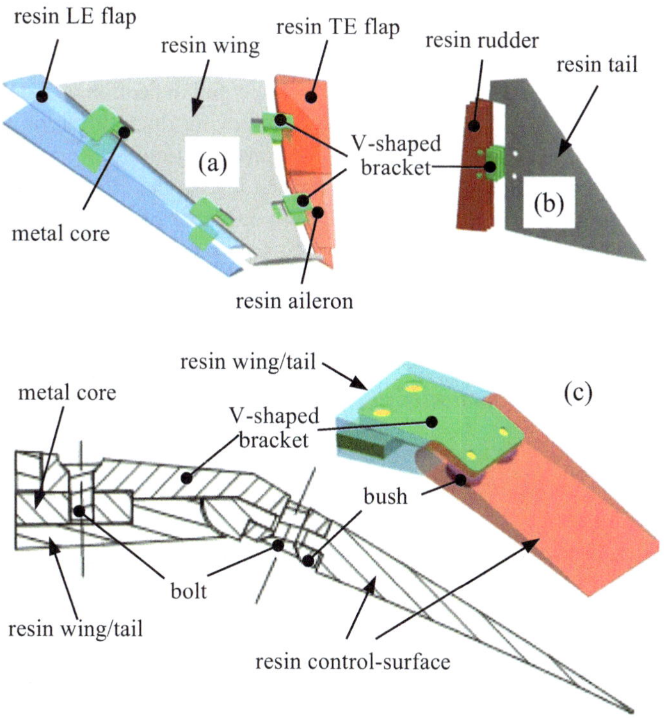

Fig. 6.5 Variable-angle piece adjustable rudder surface scheme

Fig. 6.6 Rotary axis—positioning pin adjustable rudder surface solution

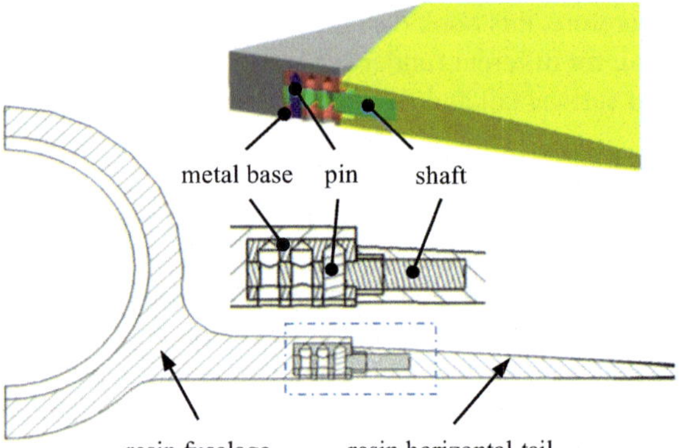

mounting holes can be machined in one piece, and no subsequent drilling is required. It should be noted that in order to ensure the installation accuracy of the resin parts during repeated disassembly, the book also reinforces the mounting holes, which will be detailed below.

2) Full-motion tail fin

For a full-motion flat tail with a small installation space airfoil, the book adopts a rotary axis-locating pin control scheme. As shown in Fig. 6.5 the metal base embedded in the resin fuselage and the metal swivel shaft in the resin flat tail are the main components to achieve the full-motion flat tail deflection, respectively. The full-motion flat tail in Fig. 6.6 is capable of achieving three different deflections, each performed by three pin holes facing different directions.

Metal parts can be machined with high accuracy, so the offset accuracy of this solution is mainly determined by the assembly accuracy of the metal base and the metal rotary shaft inside the resin matrix. This part will be discussed in the next section.

6.3.2 Resin-Metal Assembly

As discussed above, the guarantee of the overall assembly accuracy of the model depends on the assembly accuracy between metal and resin.

As Figs. 6.5 and 6.6 shown, the resin mounting holes machined by SL as a whole are the main metal-resin assembly positioning structure, and their machining accuracy has a direct impact on the assembly accuracy. According to the SL machining accuracy study, the hole machining accuracy can be improved by presetting certain compensation dimensions and optimizing the corresponding process parameters (printing direction, etc.) [72]. However, even if optimization measures are taken, the machining accuracy of holes with curved surfaces can be improved. However, even with the optimization measures, the holes with curved surfaces are still difficult to achieve the same level of machining accuracy as metals. At the same time, the model usually needs to be disassembled several times, and the resin mounting holes lose their positioning accuracy due to wear during disassembly. For this reason, this book proposes a way to install metal bushings in the resin mounting holes. On the one hand, the metal bushing can avoid the direct participation of resin in the assembly and play a protective role; on the other hand, the metal bushing can adopt an external positioning structure to avoid the loss of accuracy caused by the direct positioning of resin holes. For the size of the holes in different positions, this book proposes two mounting hole reinforcement schemes [73]. The book proposes two options for mounting hole reinforcement.

1) Direct positioning method

As Fig. 6.7a, b, the stepped hole surface of the SL machined hole is used as the positioning reference, and the metal sleeve is directly pressed into the positioning hole after applying adhesive, leaving a certain gap at the non-positioning surface to facilitate the curing of the adhesive connection. The axial positioning surface is the upper surface of the stepped hole, which is flat and can ensure high processing and positioning accuracy; while the radial positioning surface is the inner surface of the hole, which requires process optimization. This positioning method is suitable for areas with large mounting positions. The advantage is that all positioning surfaces are machined in one go by SL, which is a simple process, but the disadvantage is that the positioning accuracy depends entirely on the machining accuracy of the resin holes, which can meet the requirements in large size areas such as the fuselage, but it is difficult to guarantee the positioning accuracy of small features such as the rudder surface.

2) Auxiliary positioning method

As Fig. 6.7c, d, cylindrical pins and positioning are used to assist in the high-precision positioning and installation of the metal sleeve. The maximum length of the sleeve should be no higher than the airplane airfoil where the bonding hole is located to prevent the impact on the airfoil shape, while the sleeve is inserted on the pin to prevent the flow of bonding agent into the sleeve countersink hole. The axial and radial positioning surfaces of the metal sleeve are the positioning plate pin fit hole and the positioning plate plane respectively, and the accuracy can be guaranteed. For the positioning of the positioning plate and the resin substrate, an auxiliary positioning structure located outside the hole can be used. As Fig. 6.7d, the auxiliary positioning structure is flat, which is convenient for SL high-precision machining; therefore, this solution can ensure high positioning accuracy.

The outer surface of the sleeve is rolled mesh to increase the mosaic effect between the colloid and the surface to enhance the bonding performance; the gap between the outer surface of the sleeve and the bonding hole is taken as 1 ~ 3 mm, in order to facilitate the injection flow of the viscous resin or sufficient UV irradiation of the photosensitive resin, while also ensuring the bonding strength, the connection gap can be adjusted according to the height of the bonding hole.

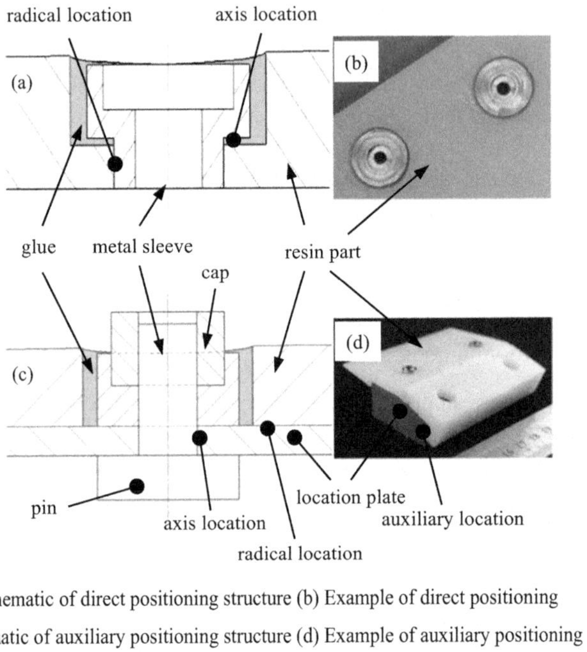

Fig. 6.7 Resin part assembly hole reinforcement scheme

6.3.3 Rudder Surface Manufacturing Accuracy

1) Print distortion

Since 3D printing technology is an ensemble of 3D entities discrete into cut layers with thickness, model layering not only destroys the continuity of the model surface, but also loses the contour information between adjacent cut layers. The thickness of the layers is the resolution of the discrete model, the larger the thickness of the layers, the lower the resolution, the more data the model loses, and the greater the printing error. The following two main types of errors are likely to occur during model printing:

(1) Truncation error

The data of the part to be machined is processed in layers, and the height direction causes loss of accuracy due to end truncation. As Fig. 6.8a, the total height of the part to be machined is H, the thickness of the stratification is $\triangle t$, and the residual height of a section of the part is $\triangle h$. When $\triangle h < \triangle t$, the height will be ignored when the data is processed discretely, thus causing a loss of accuracy.

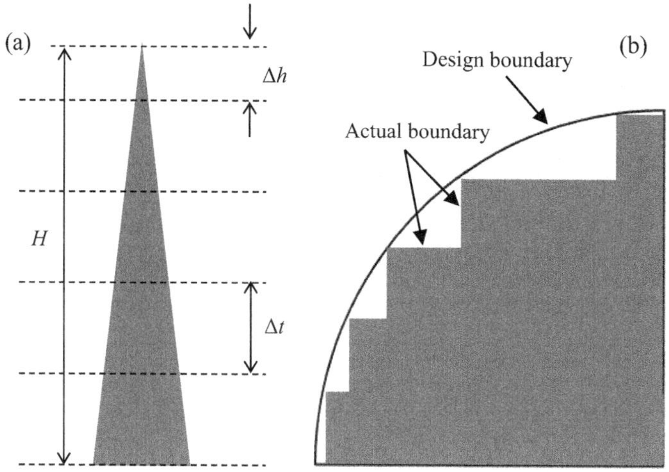

(a) Truncation error (b) Step error

Fig. 6.8 Errors caused by discrete processing of 3D printing process data

(2) Step error

When the model is layered, the shape profile of the model consists of two tangent layers, and since the horizontal surface profiles of the upper and lower tangent layers are not the same, the surface profile of the model is necessarily replaced by discrete planes, as shown in Fig. 6.8b shows, the "step effect" will be generated during processing and printing, which will significantly reduce the surface accuracy and cause surface accuracy errors.

1) Print compensation

The profile accuracy error generated by 3D printing process is related to the normal direction, radius of curvature and layered thickness of the solid surface, and the SL printing can improve the profile accuracy of the parts by reducing the layered thickness and optimizing the manufacturing direction. However, the shape of the aircraft model is complex, especially the wing part of the aircraft, and the accuracy requirements of different parts are different, so a single printing direction is not enough to meet the surface accuracy of different surfaces. Therefore, the offset compensation design can be combined with different printing directions for the parts. In addition, the resin printing process mainly relies on the polymerization chemical reaction of monomer molecules, and the resin inevitably produces volume shrinkage, especially for the holes and slots that fit with the metal skeleton, and the shrinkage size is large, so it is also necessary to design offset compensation for such structures. The offset compensation should be combined with the actual size of the skeleton to make fine adjustments appropriately.

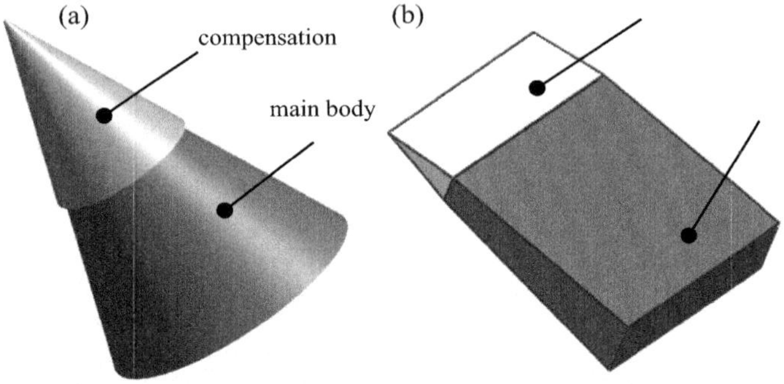

(a) Sharp and thin structure (b) Flat structure

Fig. 6.9 Dimensional compensation for printing of extreme structures

The tip of the head and other sharp and thin structure, its thin tip part is likely to cause the loss of processing characteristics due to the influence of 3D printing manufacturing process, forming a certain arc of rounded corners in the printing process, not only changes the shape characteristics of the head, but also reduces the length size of the head along the x-axis, so in the design of the tip part of the head morphological features outward offset, not only can compensate for the processing errors of 3D printing, but also can offset The amount of post-processing of the model can also offset the amount of regrinding, as shown in Fig. 6.9a shows that the front end of the head is selected by 5 mm and offset by 0.2 mm.

The flat structure such as the leading flap of the wing, due to its wing section is printed at an angle of 50°, the step effect causes the leading edge of the leading flap to have a certain dimensional contraction in the horizontal plane direction, in order to ensure the dimensional accuracy of the leading edge we offset the leading edge outward by 0.2 mm, as shown in Fig. 6.9b shows.

6.4 Case Studies

6.4.1 Model Design

The structure of a low-speed wind tunnel test model of a certain type of fighter jet is shown in Fig. 6.1. The entire aerodynamic shape is machined in SL, including the fuselage, main wing, drogue tail, flat tail, hanged bomb wing and all rudder surfaces (such as front flap, rear flap, aileron, rudder and movable flat tail, etc.). In order to improve the strength of the model and ensure the assembly accuracy, a machined metal skeleton (metal sleeve inside the fuselage and metal plate inside the main wing) is inlaid inside the resin aerodynamic shell. The mechanical property parameters of the materials used are shown in Table 1- 3 shown. The resin shell was machined

Table 6.1 Design error of center of gravity for the composite model

Coordinates	Target value	Design value	Error, %
x/mm	841.47	838.87	−0.31
y/mm	10.00	9.97	−0.30

Table 6.2 Composite model weight reduction design results

Models/parts	Metal models	Composite model—after initial design	Composite model—after optimization design
Metal skeleton/kg	69.10	27.51	26.46
Plastic housing/kg	–	15.29	9.96
All/kg	69.10	42.80	36.42
Weight loss rate/%	–	38.06	47.29

from a photosensitive resin, the wing skeleton and various connecting parts (such as the angle change piece on the rudder surface) were machined from 40Cr steel, and the material of the other skeletons was 45 steel. The test wind tunnel used in this book is FD-09 low-speed wind tunnel of China Aerospace Aerodynamic Technology Research Institute, with the test section size of 3 m × 3 m, the model size scaling factor of this book is 1:10, and the length and span of the fuselage are about 1.5 m and 0.9 m. The machining size of SL equipment used in this book is 0.60 m × 0.60 m × 0.45 m (SPS600B, Xi'an Jiaotong University). Therefore the resin shell needs to be split for processing. At the same time, this division also provides the possibility of assembling the model. As shown in Fig. 6.1, the fuselage is divided to avoid important aerodynamic parts, and the wing and fuselage are machined in one piece (left and right main wing—middle fuselage and drogue—rear fuselage) to avoid aerodynamic interference between the wing and fuselage joints. All the rudder surfaces and hangers are machined and mounted separately.

The position of the center of gravity of the model after optimization design is shown in Table 6.1 The design accuracy is good and its error can be neglected.

The weights of the optimally designed models are shown in Table 6.2. The weight reduction ratio of the optimally designed composite model is nearly 50% compared to the all-metal model, i.e., the weight is reduced by half.

6.4.2 Model Calibration

To ensure the safety and effectiveness of the wind tunnel test, the load bearing and vibration conditions of the model need to be calibrated before the test. Among them, the load bearing calibration is to examine whether the strength and stiffness of the model meet the requirements under the extreme test conditions, which is carried out by the combined Computational Fluid Dynamics (CFD)—Computational

Structure Dynamics (CSD) method [74]. The vibration calibration is performed to ensure that the model-computerized structural dynamics is not too rigid. The vibration calibration is to ensure that the first-order inherent frequency of the model-balance-strut system avoids the peak blowing frequency (related to the wind tunnel and working conditions, provided by the wind tunnel test department), and the inherent frequency of the model-balance-strut system is calculated by the Finite Element Analysis (FEA) method.

(1) Pneumatic Analysis

Because the aircraft is symmetric relative to the fuselage center, a calibrated half-model can be used to reduce the computational effort. In this project, the NS equation for 3-D Reynolds averaged non-compressible flow (3-D Reynolds averaged non-compressible Navier–Stokes equation) is used to model the CFD analysis, and the discrete modeling method is the second-order Finite Volume Method (FVM) [75–78], a k-ω turbulence model is used. In the flow-solid cross section, a non-slip boundary condition is used, where the boundary is free of heat transfer and adiabatic treatment is done as follows [79].

$$u = v = \omega = 0, \partial P/\partial \boldsymbol{n} = 0\, and\, T = T_{a\omega} \tag{6.12}$$

The CFD flow field mesh is a tetrahedral unstructured mesh, with the total number of cells in the order of 1 million, as shown in Fig. 6.10. The input parameters for the pneumatic analysis are shown in Table 6.3. The incoming static pressure and reference temperature are taken from the wind tunnel data, and the fluid velocity and angle of attack are extreme conditions for this test.

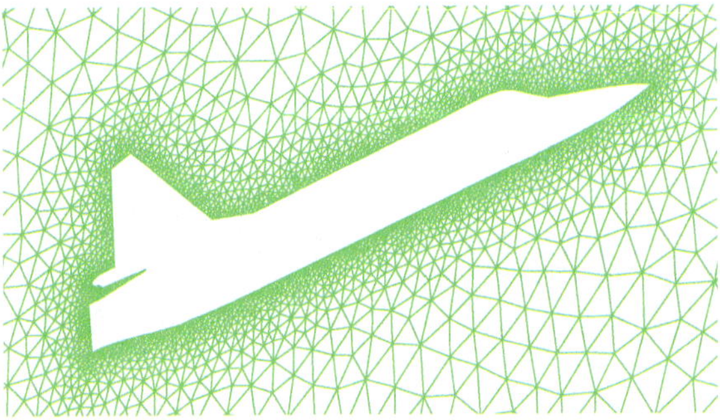

Fig. 6.10 Aerodynamic mesh in CFD analysis

Table 6.3 Pneumatic analysis parameters

Pneumatic parameters	Value
Incoming air pressure,P /atm	1
Reference temperature,T /K	300
Flow field velocity v, /m $\cdot$ s^{-1}	50
Angle of attack, α/°	8

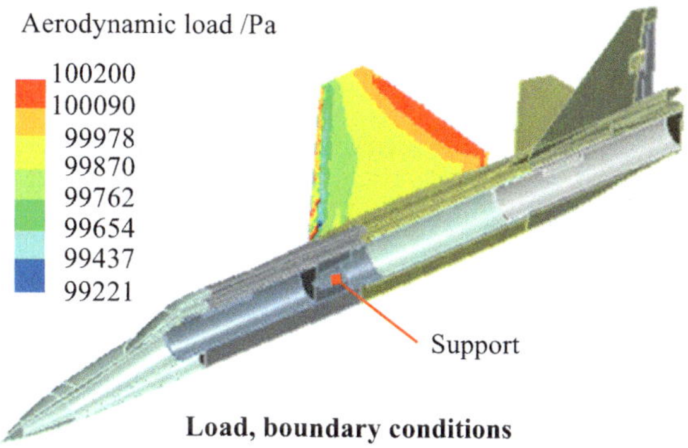

Fig. 6.11 CFD calculation of the pressure distribution of the flow field on the lower surface of the wing

Figure 6.11 shows the pressure and velocity distribution of the flow field on the lower surface of the model wing obtained by CFD calculation, as expected. The pneumatic loads will be imported into the subsequent structural analysis for load calibration.

(2) Load-bearing calibration

The purpose of the load-bearing calibration is to calibrate the strength and stiffness performance of the model under extreme aerodynamic loads, which are solved using the CSD method with the aerodynamic forces calculated by CFD as the load. The mathematical model is shown in the following equation [80, 81], the discretization method is the finite element method. where $[M]$, $[C]$ and $[K]$ represent the mass, damping and stiffness matrices of the model, respectively. $F(t)$ In the load function, transformed from CFD analysis, the $q(t)$ is the displacement vector with the solution.

$$[M]\{\ddot{q}(t)\} + [C]\{\dot{q}(t)\} + [K]\{q(t)\} = F(t) \tag{6.13}$$

Four extreme conditions were chosen for the calibration. As shown in Table 6.4, the blowing speed and angle of attack are the maximum values in all conditions, 60 m/s and 8°. Condition 1 calibrates the overall strength and stiffness of the model with all control surfaces in the initial position; Condition 2, Condition 3 and Condition 4 calibrate the strength and stiffness of the rudder, front flap and rear flap/aileron, respectively, with the corresponding control surfaces in the maximum open position.

The calculated stress and deformation diagrams of the model (including metal and resin parts) are shown in Fig. 6.12. At the safety factor of 2, the allowable strengths of the steel and resin materials used in the model are 194 MPa and 15.6 MPa, respectively, and the maximum stresses of the model under the extreme conditions used are less than the allowable values, and the strengths are safe.

(3) Vibration calibration

In order to avoid resonance damage to the model during the blowing process, the first-order intrinsic frequency of the model-balance-strut system needs to be calibrated to

Table 6.4 Wind tunnel test conditions of the composite model

Status		Model status			Control surface status		
Number	Testing	Angle of Attack [a] α /°	Speed v/ $m \cdot s^{-1}$	Side Slip Angle β/°	Front flap deflection angle δlef/°	Back flap deflection angle δf/°	Steering rudder δr /°
1	Cruise	A10	50, 60, 70	0	0	0	0
2	Side Slip Angle	A10	50	−2, 2, 4	0	0	0
3	Front flap	A10	50	0	24	0	0
4	Backlap	A10	50	0	24	25	0
5	Steering rudder	A10	50	0	0	0	15, 25

[a]Angle of attack variation range: A10 = −4, −2, −1, 0, 1, 2, 4, 6, 7, 8°

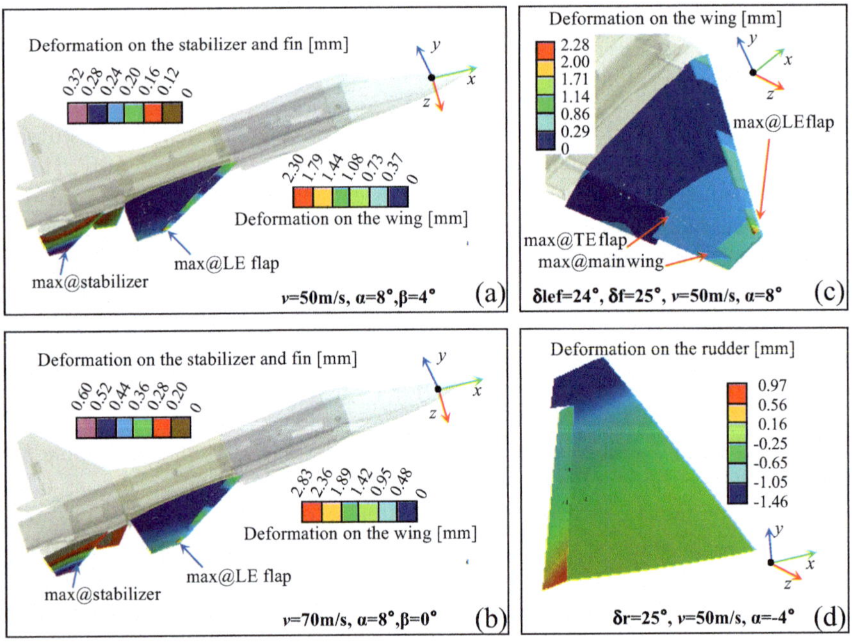

Fig. 6.12 Stress and deformation distribution of model metal and resin parts

ensure that it avoids the extreme value of the blowing pulsation frequency of the wind tunnel [82]. The frequency value of FD-09 wind tunnel is about 10 Hz (at the ultimate wind speed of 100 m/s), and the design requires the first-order intrinsic frequency of the model-balance-strut system to be larger than this value. The inherent frequency of the system is determined by its stiffness, damping and mass properties [80]. Under

Table 6.5 First-order eigenfrequencies of different models

Models	First order frequency, Hz
Metal models	10
Composite model—after initial design	16
Composite model—after optimization design	21

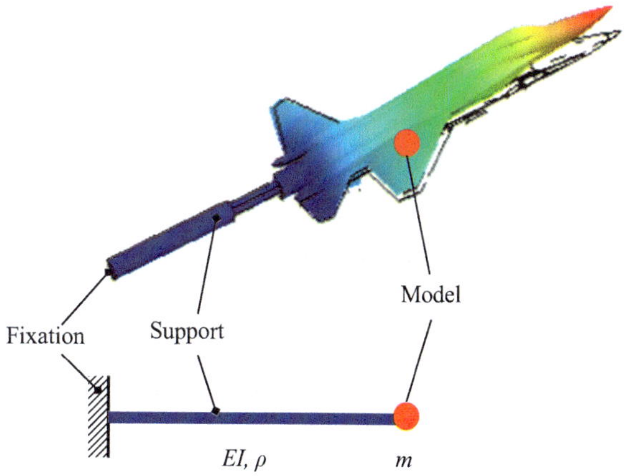

Fig. 6.13 First-order inherent frequency of the model-balance-strut system

the premise of the first two parameters, the lower the mass of the system, the higher the first-order inherent frequency and the better the corresponding vibration safety.

As shown in Table 6.5, the first-order intrinsic frequencies of the pure metal model, the initial composite model and the optimized composite model (with optimized weight reduction design) increase in order, and the optimized model has a maximum value of 21 Hz. The first-order intrinsic modes of the model-balance-strut system are shown in Fig. 6.13, which is the up-and-down oscillation around the strut along the y-axis, and its determinants are strut stiffness, strut damping and system mass. The first two parameters are the same for all three models, and the weight of the model is the determining factor. The optimized model has a smaller mass (about 50% weight reduction) compared to the pure metal model, so its first-order intrinsic frequency is larger. This value is much larger than the extreme frequency of wind tunnel fluctuations (10 Hz), so the vibration safety of the composite lightweight model is good.

6.4.3 Wind Tunnel Test

The machined and assembled model is mounted on the wind tunnel balance strut, as shown in Fig. 6.14. Compared to the model built in this book, the metal model used as a control has two differences: (1), there are artificial turning strips affixed

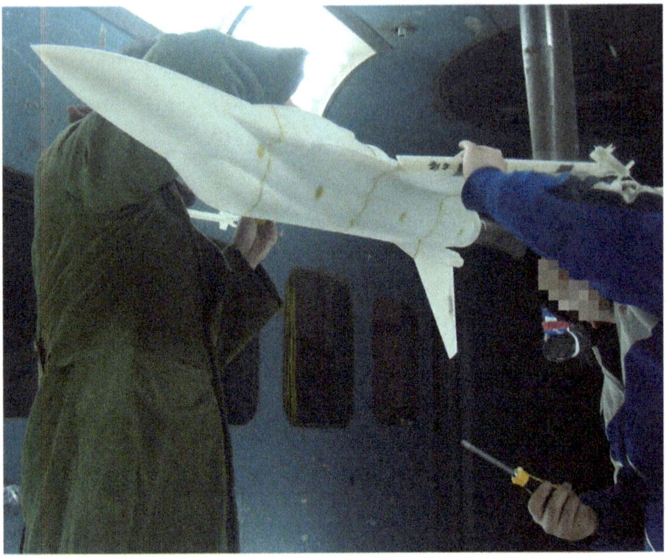

Fig. 6.14 Model installed in FD-09 wind tunnel

to the nose and leading edge of the flap; (2), the air intake is naturally vented. The deviations in wind tunnel test results due to the structural differences are discussed in the next section.

The test conditions were arranged according to the optimal design method, shown in Table 6.4. The wind tunnel tests tested the effects of angle of attack, blowing speed, model sideslip angle, and deformation angle of each control surface on the aerodynamic parameters of the model.

6.5 Analysis and Discussion

6.5.1 Contrast Analysis

This section discusses the economics of model design and fabrication, focusing on the model's part overview, total weight, machining cycle time, and machining costs. A comparison of each metric for composite and metal models is shown in Table 6.6.

(1) Total number of parts

As shown in Table 6.6, the total number of parts of the composite model is reduced by about 50% compared to the metal model, mainly due to the integrated manufacturing of the SL process. The significant reduction in the number of model parts contributes to shorter model machining cycles and has positive implications for the

Table 6.6 Comparison of the economics of different processing methods

Projects	Model of this book	Metal models	Ratio, %
Number of parts	31	60	52
Weight/kg	36.84	69.90	53
Processing cycle/h	264	1500	12
Processing cost/RMB	¥250,000	¥1,000,000	25

overall model reliability; fewer mounting surfaces reduce the need for mounting parts, which means lower total model weight; and the integrated shape reduces data fluctuations caused by aerodynamic clearances, which has obvious implications for improving data reliability.

(2) Total weight of the model

The total weight of the composite model is reduced by about 50% compared to the metal model, as Table 6.6 shows. In addition to benefiting from the reduction in the total number of parts, as mentioned earlier, there are two other determining factors: on the one hand, the composite model uses resin instead of metal to compose the aerodynamic profile; On the other hand, the weight reduction optimization design removes mass and has an important impact on the weight reduction of the composite model, both of which depend on the entry of the SL process.

(3) Processing cycle and cost

The improvement in model processing cycle time and cost is even more pronounced, as Table 6.6 shows, these two parameters of the composite model are only 12% and 25% of those of the metallic model. On the one hand, the resin pneumatic shape machined by SL is directly influenced by SL, which can greatly reduce the cycle time and cost; on the other hand, the structure of the machined metal part can be greatly simplified because it only provides strength and stiffness in the inner cavity, and its machining cycle time and cost can be optimized. In addition, the design efficiency of the model is also simplified. Although it is difficult to analyze quantitatively, experience shows that after the introduction of SL technology, the constraints of machining process on the model structure are greatly reduced, especially SL has better machining capability for complex internal and external structures, and designers can rely on optimal design tools to select the internal and external structures with optimal performance, and the difficulty of model design is greatly reduced [83, 84]. The difficulty of model design is greatly reduced.

From the above analysis, the composite model based on SL processing has a better economy, especially in terms of total part weight reduction and processing cycle time, which reflects the characteristics of 3D printing technology represented by SL. Compared with the metal model, the total weight of the SL composite model is reduced by about 50%, achieving further lightweighting of the force measurement model, which is unachievable with conventional processing technology, which has positive implications for the preparation, transportation, and testing safety of the model. Meanwhile, the comparison test shows that the composite model with full SL resin profile in the basic state obtains well-matched data, especially the basic lift coefficient curve $C_L - \alpha$ that almost overlap with that obtained from the metal model with high accuracy. This shows that the model obtained by the newly implemented method can be used for the basic test items.

6.5.2 Technical Limitations

Although the model implementation based on SL technology has the above advantages and reflects the high value of application, there are still areas for improvement. On the one hand, the above conclusions come from case studies, and their applicability needs to be discussed in detail in order to extend the conclusions drawn from individual objects to a wider range; on the other hand, the technical solutions proposed in this book also have certain applicability targets, and the applicability to general wind tunnel test models needs to be further discussed.

1) Subjects of the case study

The case study object is a single-engine light fighter with a length of 14.97 m, a wingspan of 9.46 m, and a height of 4.78 m. It adopts a normal aerodynamic layout with a medium delta-wing span-to-edge ratio, a maximum takeoff weight of 12,474 kg, a maximum aircraft speed of Ma1.8, and a maximum overload of about 8g, which is less loaded than other types of aircraft (heavy fighters, transport aircraft, etc.). The model in this chapter is the full-engine force measurement model of the fighter, and the air inlet is simplified as a conical block, and all rudder surfaces are adjustable; the test wind tunnel used is the FD-09 low-speed wind tunnel of the China Aerospace Aerodynamic Technology Research Institute, with a test section size of 3 m × 3 m, compared with other low-speed wind tunnels (e.g., FL-13 test section of 8 m × 6 m/16 m × 12 m), the size of the model used in this wind tunnel is smaller. The size of the model is smaller than other low-speed wind tunnels (e.g., FL-13 test section is 8 m × 6 m/16 m × 12 m). In this book, the blowing range in the low speed stage (up to 70 m/s), angle of attack and side slip angle adjustment range are small, and the rudder adjustment range is normal. Therefore, the object of study in this book is representative of smaller models, control surface effects, and low speed tests, and special verification should be done for the extension of large load prototypes, large size models, large angle of attack ranges, and high speed blowing tests.

2) Breadth of data tested

This book has done sufficient wind tunnel tests on the composite model, including basic cruise test, sideslip deflection angle test, front flap efficiency test, rear flap efficiency test and rudder efficiency test, and analyzed the experimental results; in order to study the accuracy of the data obtained from the composite model, the data from the basic cruise were compared with the metal model, and preliminary conclusions were drawn. However, the test data in this book still need to be further expanded: (1) full-motion flat-tail efficiency test is not conducted in this book, which needs to be supplemented; (2) the qualitative analysis of all the results is only trended, which helps to get the preliminary results but is slightly insufficient for more refined analysis and needs to be improved; (3) the comparison data of the metal model is not sufficient, especially the lack of comparison data in rudder efficiency, which needs to be further expansion.

3) Problems with all resin profiles

Through the strength check test, it can be seen that the designed model scheme can meet the test needs of this study, but the results show that the resin strength margin is small and the model deformation is large. Further analysis shows that the main part of the problem is on the rudder surface, especially the aileron and rear flap, which are subject to large forces. For the object and test range proposed in this book, the all resin rudder surface is safe and reliable for aerodynamic data collection in the basic no-deflection state, but it will be a serious problem for aircraft models with rudder deflection, large angle of attack, high speed test and other high load classes. Therefore this part is not suitable for SL resin processing and needs to be realized by means of metal mechanical processing.

The advantage of the practical lightweight model implementation method proposed in this book is the lightness of the model and the economy of manufacturing, so the impact on the overall effect of the technology when this part is machined from metal instead needs to be discussed. The total number of parts and the difficulty of assembly remain unchanged for both options, and the impact on the weight and machining cycle/cost of the model needs to be considered. As Table 6.7 shows, the overall weight gain of the model after the rudder surface (including the tailplane, etc.) is changed to machined by aluminum alloy or steel is 0.91 kg and 3.67 kg, respectively. According to Table 6.3, the total weight of the composite model after optimization is 36.42 kg. Even if all the control surfaces (including the tailplane) are made of 40Cr steel, the weight increase is only about 10%, and the weight reduction is still about 40% compared with the all-metal model, so the advantage of light weight is still obvious. It can be seen that changing the rudder surface to metal processing has less effect on the lightweight effect of the composite model, and this effect may be further reduced if it is considered that only part of the high-load rudder surface (such as rear flap and aileron) can be replaced by metal while retaining the other rudder surface materials as SL resin according to the test load. Similarly, the cycle time and cost of composite model processing may be increased after the rudder surfaces (some or all) are mechanically machined from metal, but still maintain some advantages over the all-metal model implementation scheme.

Therefore, making full use of the advantages of SL process and other processing processes and SL resin materials and metal materials, integrating the advantages of the two model realization methods, and based on the research in this book, a model realization scheme that can expand the scope of application of the technology is proposed: (1) maintaining the main structure of resin shell-metal skeleton unchanged; (2) SL resin in the parts of the fuselage, main wing structure and other parts with larger cross-sectional dimensions, with higher strength margin, with greater expandability,

Table 6.7 Effect of variation of model rudder surface material on mass

Mass /kg	Backlap	Aileron	Steering rudder	Front flap	Horizontal tail fin	Vertical tail fin	Total weight
Resin-14120	0.094	0.026	0.014	0.155	0.143	0.187	0.619
Aluminum-7A04	0.234	0.064	0.033	0.384	0.356	0.461	1.532
Steel-40 Cr	0.654	0.179	0.093	1.074	0.996	1.291	4.287

and it is expected that similar parts of the model used in high-load aerodynamic layout and high-speed tests can be processed using SL resin; (3) for the weaker edge parts, change to the realization of high-strength metal mechanical processing (such as each rudder surface), especially the ailerons and rear flaps, which are subject to more serious forces. The specific replacement range is determined by different experimental objects and test ranges, according to the strength calibration results. This new scheme has no effect on the model design, total number of model parts, and assembly process, and has less effect on the total weight of the model and model machining periodicity and cost, basically retaining the advantages of the technical scheme of full SL resin aerodynamic shape, but making the composite lightweight model's expected to be extended to other high load tests.

Chapter 7
Additive Manufacturing of Wind Tunnel Test Models for Pressure Measurement

7.1 Overview

7.1.1 Pressure Measurement Methods of Wind Tunnel Test Model

The main purpose of the pressure measurement test is to provide the original data of the aerodynamic load distribution for the strength calculation of the aircraft and its components, to provide a test basis for the study of the aerodynamic performance of the aircraft and its components, to study the flow characteristics around the model, and to determine the location of the minimum pressure point on the wing, the airflow separation characteristics, as well as the lift force acting on the model, the differential pressure resistance and the location of the center of pressure through the pressure distribution measurement. Information [3] The most commonly used pressure measurement systems are to determine the location of the minimum pressure point on the wing, the airflow separation characteristics, and the lift, differential drag, and center of pressure acting on the model. The most commonly used pressure measurement system is to build a special pressure measurement model, make a pressure measurement hole along the surface normal to the location of the pressure measurement point, and connect it to a measurement device such as a manometer or pressure sensor through a pressure measurement tube. Not only the whole system is very complicated, causing some interference to the flow field around the model, but also the processing of data is very cumbersome, so later the pressure sensor with pressure conversion valve was introduced, and the situation has been improved to some extent.

Russia, the United States and Europe have been working on optical pressure distribution measurement methods and techniques based on Pressure Sensitive Paint (PSP) since the 1990s. This technique is a revolutionary advance compared to the traditional pressure sensor-based pressure measurement techniques. The surface pressure of the

W. Zhu and D. Li, *Models for Wind Tunnel Tests Based on Additive Manufacturing Technology*, Additive Manufacturing Technology,
https://doi.org/10.1007/978-981-99-5877-1_7

measured area and its distribution are measured optically by applying a pressure-sensitive paint to the surface of the measured model, and its fluorescence intensity varies with the airflow pressure. This non-contact measurement method is flexible, easy to use, and offers significant cost and time savings. The measurement of pressure distribution over the entire surface of the model eliminates the need to create pressure holes, pressure pipes, and wiring, saving the expensive cost of manufacturing dedicated pressure measurement models. More importantly, the pressure data obtained is continuous and extensive, especially in areas where it is not possible to install a pressure line, and can be measured by this method [85]. The pressure can be measured especially in the area where it is not possible to install a pressure line. In low-speed wind tunnels, the pressure-sensitive coating PSP technique suffers from testing accuracy problems because the pressure on the model surface is at a low level. Therefore, for the low-speed wind tunnel test with wind speed less than Mach 0.4, the optical pressure distribution measurement method based on PSP has not been able to be applied, mainly because of the weak luminescence intensity of the pressure-sensitive coating in the low-speed low-pressure blowing wind test, which brings great difficulties to the optical detection and image data processing, and it is difficult to meet the test requirements. On the current level, the optical pressure measurement method is still only suitable for cross, supersonic wind tunnel test. Specifically in China, the technology to reach the level of being able to completely replace the open-hole pressure measurement there is still a lot of work to be done, such as the sensitivity, stability and life of domestic coatings and other performance, the development of multi-component coatings, film-forming technology, data acquisition and processing methods and calibration methods and other technical issues, each of which is a great research work. Therefore, there is still a way to go for the engineering application of this technology. In summary optical pressure measurement has a broad application prospect, but at present in the low-speed wind tunnel test still can only take the open-hole pressure measurement method.

7.1.2 Requirements for Pressure Measurement

The selection of model pressure profiles and pressure points is based on the test task and the actual possible arrangement of the model. For the wing, in general, two to seven binary airfoil pressure measurement profiles can be taken, and three to five ternary airfoil pressure measurement profiles are generally taken. In addition, the pressure measurement points along the chord direction on the upper and lower surfaces of the wing profile should be no less than fifteen, which can usually be arranged at 0, 1.25, 2.5, 5, 10, 15, 20, 30, 40, 50, 60, 70, 80, 90, 95 and 100% of these chord direction positions on the upper and lower surfaces of the wing. The number of pressure measurement points can be appropriately increased in areas with more drastic pressure changes such as the leading edge attachment of the wing and other locations. If a large-scale pressure test is conducted, the requirements for measurement equipment and measurement technology will be higher, for example, in a 2.4m wind tunnel, the number of pressure measurement points can be completed in thousands of pressure tests [86]. For example, thousands of pressure measurement points can be completed in a 2.4m wind tunnel.

The pressure measurement tube inside the pressure measurement model is generally made of purple copper tube with an inner diameter of 0.4~1.2 mm or stainless copper tube with annealing treatment. The requirements for the pressure measurement piping are as follows:

(1) The length of the pressure measurement tube should be as short as possible, if it has to be led out to the outside of the model, plastic pipe can be used to connect, the plastic pipe in the model should be dragged out of the position to consider the two aspects of the small bypass interference and easy installation;
(2) The pressure measurement tube requires a clear direction, orderly arrangement, and accurate identification with the corresponding serial number of the pressure measurement hole;
(3) The gas tightness and aeration of all pressure measurement piping must be carefully checked before the experiment;
(4) When processing the pressure measurement orifice, avoid right-angle bends in the pressure measurement tube to prevent breakage and pipe blockage;
(5) The axis of the pressure measuring hole should be perpendicular to the surface of the model, requiring no chamfering of the orifice and smooth, no burrs or unevenness around the orifice.

7.1.3 Comparison of Model Manufacturing Methods for Pressure Measurements

From the conventional processing methods of pressure measurement orifices can be seen, pressure measurement model orifice structure is complex, the number of traditional mechanical processing methods is very difficult, a pressure measurement model processing costs may be up to a million yuasn(RMB), taking more than six months. Light-curing 3D printing technology has the characteristics of complex structure easy to print, very suitable for the intricate orifice structure on the pressure measurement model, can realize the pressure measurement hole, pressure transmission pipeline and model shape integrated printing. In addition, light-curing 3D printing technology also has and has the advantages of high speed, high precision, good surface quality, simple and time-saving processing.

From the above analysis of the structure of the conventional metal model, it is clear that the 3D printing manufacturing method has incomparable superiority to the conventional manufacturing method in the manufacturing of the pressure measurement model, so it has a very broad application prospect. The advantages and disadvantages of the conventional metal model using CNC machining method and 3D printing manufacturing method are shown in Table 7.1.

Table 7.1 Advantages and disadvantages of the manufacturing method of the wind tunnel test model for pressure measurement

Model manufacturing methods	Advantages	Disadvantages
CNC machining of metal models	High precision, high strength, high stiffness	Complex process, long manufacturing cycle, high cost, difficult to process internal structure
3D printed plastic models	Integrated shape and aperture printing, fast and low cost; easy to print thin flaps, ailerons, rudder surface pressure measurement apertures	Slightly less accurate but can meet wind tunnel pressure measurement requirements

7.2 Effect of Structure Morphology on Pressure Measurements

7.2.1 Influence of Pressure Measurement Piping Parameters on Pressure Measurement Results

The structural parameters of the pressure measurement pipeline have a close relationship with the wind tunnel pressure measurement experiment error, mainly in the diameter of the pressure measurement hole, the degree of pipe bending, pipe length and cross-sectional area of three aspects.

1) The diameter of the pressure measuring hole

When measuring the pressure and its distribution on the surface of a wind tunnel test model by a pressure measurement hole, the presence of the hole changes the curvature of the flow line and the vortex inside the hole near the location on the model surface, which results in a higher pressure value P_m measured by the hole than the real pressure value P_t. The pressure measurement error is generally characterized by $\Delta P/q$, where $\Delta P = P_m - P_t$, and q is the dynamic pressure. Ideally, a pressure measurement orifice with zero orifice diameter will not cause any disturbance to the air flow in the vicinity of the pressure measurement orifice. As shown in Fig. 7.1, the pressure measurement error $\Delta P/q$ caused by the finite size pressure orifice increases with the diameter of the orifice on the one hand, and with the number of airflow *Ma* on the other hand [87]. The pressure measurement error ΔP/q caused by the finite size pressure orifice increases with the diameter of the orifice on the one hand and with the number of Ma of the airflow on the other. Assuming a pressure hole diameter *d* of 1 mm and airflow Mach numbers *Ma* of 0.4 and 0.8, the pressure measurement errors $\Delta P/q$ caused by the pressure hole are 0.6% and 1%, respectively. Therefore, from the aspect of reducing the influence of the pressure measurement hole on the pressure measurement results,

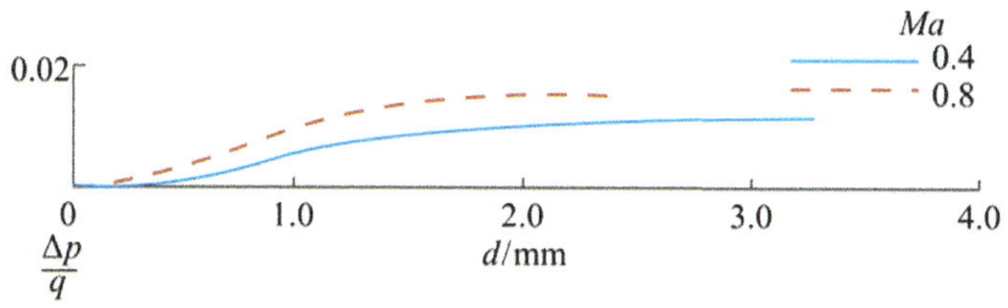

Fig. 7.1 Effect of bore size on pressure measurement error [87]

the pressure measurement hole with small orifice diameter should be designed as much as possible. The ability of light-cured 3D printing technology to fabricate miniature orifices will be investigated in Chap. 2.

2) The degree of pipe bending

In general, to ensure the stability of the airflow at this location, the vertical depth of the pressure measurement hole should be greater than twice the diameter of the orifice. When pressure is transmitted in the pressure transfer flow path, the bend in the flow path will have a restraining effect on the airflow. The greater the bend, the greater the resistance, which results in a longer time to reach pressure equilibrium. The calculation of the drag coefficient of a bent pipe is related to the Reynolds number *Re*, the diameter *d* of the bend and the radius of curvature *R of the* axis of the bend, and the Dean number *De is* usually used to characterize this effect in a comprehensive manner, $De = Re\sqrt{d/R}$. Therefore, in order to avoid causing sharp bending and increase the flow resistance, the design of the orifice should minimize the number of turns in the flow channel and the degree of bending of the elbow, such as bending angle as far as possible to bend into obtuse angle, avoid acute angle, etc.. In addition, in order to facilitate the cleanup of the residual liquid resin in the conduction 3D printing orifice, the bending transition connection line and transition cross-section can not be designed as angular, available arc connection, elliptical arc connection and other curvature light smooth connection, and with different diameters of the circular cross-section for sweeping transition.

3) Pipe length and cross-section size

Because the pressure measurement pipeline is long, in addition to the pressure measurement orifice installed inside the model, an external pressure pipeline must be installed to lead the airflow out of the wind tunnel to connect with the pressure sensor, thus leading to a large hysteresis effect of the pressure measurement instrument. Therefore, when connecting the pressure sensor and scanning valve, it should be placed as close to the pressure measurement point as possible, in order to shorten the length of the pressure measurement pipe, thus reducing the pressure hysteresis caused by the pressure transfer, the pressure measurement error caused by the pressure hysteresis can be calculated by the following formula [87] Calculation:

$$\left(p_t^2 - p_m^2\right)\left(\frac{A^2}{16\pi\mu l}\right) = \frac{dp_t}{dt}\left(\frac{Al}{6}\right) + \frac{dp_m}{dt}\left(V_m + \frac{Al}{3}\right) \tag{7.1}$$

Eq

p_t the actual pressure at the pressure measurement;
p_m pressure recorded by the sensor;
A cross-sectional area of the manometer tube;
l The cross-sectional length of the pressure measuring tube;
μ the coefficient of viscosity of the gaseous medium;
V_m sensor gas chamber volume.

Suppose $V_m << Al/3$, then: $\frac{dp_t}{dt} \approx \frac{dp_m}{dt} = \frac{dp}{dt}$; and $p_t \approx p_m = p$, then Eq. (7.1) can be simplified as

$$\Delta p = \frac{4\pi\,\mu\,l^2}{Ap}\frac{dp}{dt} \tag{7.2}$$

In Fig. 7.2, the curve of $\Delta p/p - dp/dt$ for the pressure measuring line is given. When $dp/dt = 50$ Pa/s, the diameter of the pressure measuring hole $d = 0.51$ mm, and the pipe length $l = 0.91$ m, $\Delta p/p = 0.06\%$; and when increasing l by a factor of 5, $l = 4.57$ m, $\Delta p/p = 1.5\%$. In order to eliminate the pressure measurement error caused by the pressure transfer lag, wind tunnel experiments are generally used to increase the pressure stabilization time and wait until $dp/dt \approx 0$, i.e., $P_m \approx P_t$, before starting to collect experimental data [87] It can be seen that shortening the pipeline and increasing the pipe size can reduce the pressure measurement error. It can be seen that shortening the pipeline and increasing the cross-sectional area of the pipe can shorten the pressure stabilization time and thus save the blowing time.

In addition, the pressure state before and after the pressure reaches equilibrium at the pressure measurement points at different locations during the blowing process is different, and the lag time grows as the pressure goes from high to low pressure. When the pressure at the pressure measuring hole is higher than the pressure at the measuring instrument, the air density in the pipe increases until it reaches the equilibrium state.

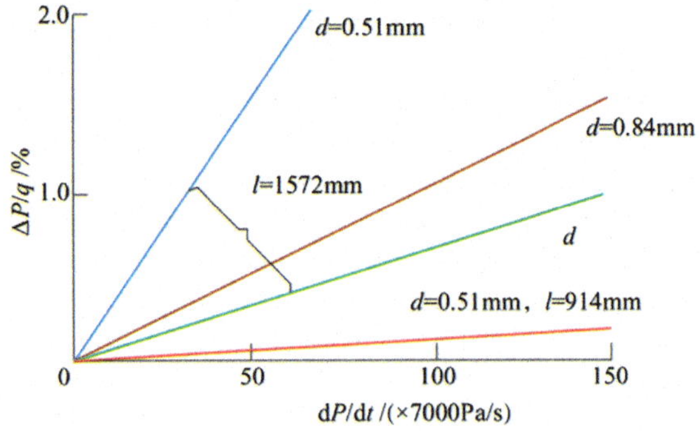

Fig. 7.2 Pressure measurement error due to pressure transfer lag [87]

As the density increases, the specific volume of air in the pipe decreases, resulting in a decrease in the velocity of airflow in the pipe, so the pressure loss in the pipe is small when it is close to the equilibrium state. When the pressure at the pressure measuring orifice is lower than the pressure at the measuring instrument, the specific volume of air inside the tube increases, resulting in an increase in the velocity of the airflow, so that the pressure loss near the equilibrium state is large. If a sensor is used inside the model as well as a scanning valve device, the required gauge tube length is correspondingly reduced, and therefore the pressure lag time is also shortened. However, it still takes a few seconds for stabilization when the scanning valve is switched from one gauge orifice to another to record the data. When arranging the sensor inside the body, it is important to ensure that the pressure sensor and its entire accessory tube system are easy to install and maintain, and that the sensor is isolated from the outside environment, such as vibration and heat, so that it does not affect the sensor's performance.

7.2.2 Influence of Orifice Printing Defects on Pressure Measurement Results

Static pressure is the pressure acting in the direction normal to the wall of the flow tube during the flow of fluid. The local static pressure is felt by opening a small hole along the normal direction on the wall or model surface of the test section of the wind tunnel. If the pressure is measured after a long enough period of pressure equilibrium stabilization, the effect of the error mainly comes from the quality of the print at the pressure measurement hole. The shape of the pressure hole edge, the inclination of the hole axis, bulges (or depressions) near the hole, scratches, etc. all have an effect on the pressure measurement value [87]. When the pressure hole edge is rounded, part of the dynamic pressure will be felt, and the pressure measurement value is high; when the pressure hole edge is chamfered to form a conical flare, there will be airflow separation, and the pressure measurement level is low, such as Fig. 7.3. When the axis of the pressure orifice is tilted toward the incoming flow, the pressure measurement value increases; when the axis of the pressure orifice is tilted toward the downstream, the pressure measurement value decreases, as shown in Fig. 7.4. When the back of the pressure hole protrudes, the measured value is high; when the front of the pressure hole protrudes, the measured value is low; when there are burrs near the pressure hole, the measured value is low; when there are uneven corrugations on the surface near the pressure hole, it makes the measured value decrease, as shown in Fig. 7.5.

The step effect and surface treatment have to be removed after 3D printing. Due to the step effect in the 3D printing principle, the step formation direction of the orifice and vertical hole wall of the pressure measurement orifice is different in different printing directions. When the surface of the measurement model is not polished, there must be a certain protrusion near the orifice due to the 'step effect', and the

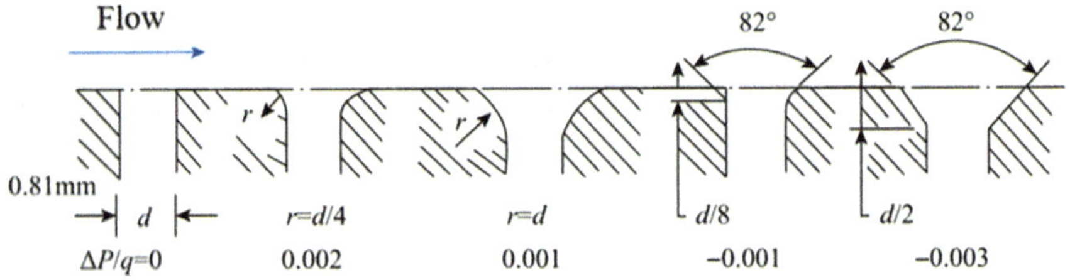

Fig. 7.3 Error of pressure measurement value caused by irregularity of hole edge [87]

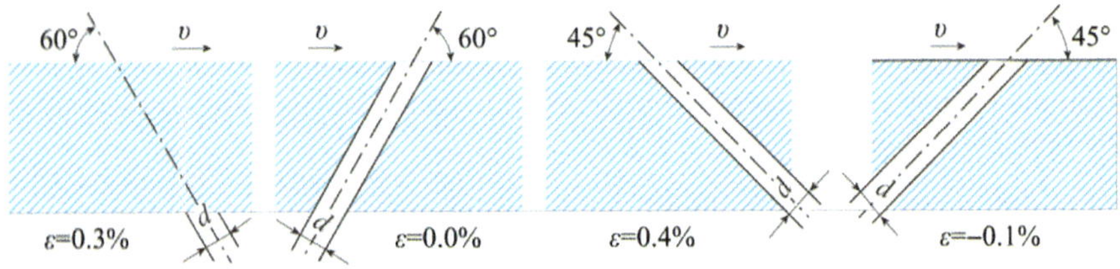

Fig. 7.4 Error in pressure measurement value due to hole axis deflection [3]

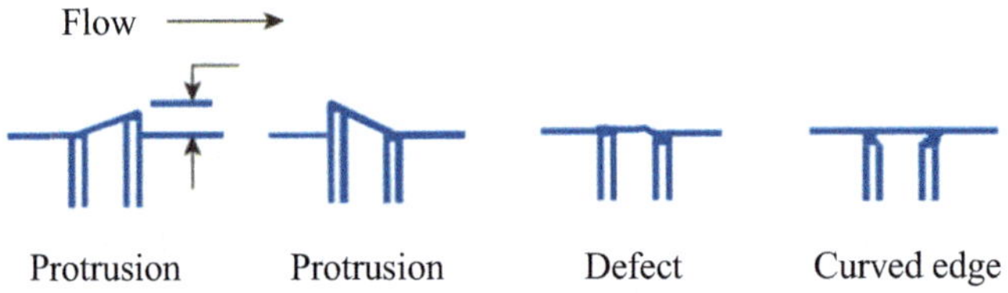

Fig. 7.5 Defects such as protrusions and scratches near pressure holes [87]

edge of the orifice may be scratched during the polishing process. Even if the surface is properly treated, it is still impossible to form a regular round hole edge because of the step effect of the hole wall. A square flow field model was established to simulate the wind tunnel wall, and a 0.8 mm pressure measurement hole was opened in the center of one wall of the square flow field, and the simulation was carried out in a CFD software. Figures 7.6 and 7.7 shows the pressure of printed holes with 90° vertical and 0° horizontal printing directions. The flow field inlet velocity V is 100 m/s. It can be seen that different hole types have different effects on the airflow near the pressure measurement hole.

The flow rate and pressure at the pressure measurement orifice and the pressure at the measurement point are shown in Table 7.2. The pressure measurement error for the step orifice compared to the ideal circular orifice is calculated schematically as follows (ρ is the density of the incoming gas)

$$\varepsilon = \Delta p/q = \frac{\Delta p}{\frac{1}{2}\rho V^2} = \frac{351.91 - 411.52}{\frac{1}{2} \times 1.29 \times 100^2} = -0.92\% \tag{7.3}$$

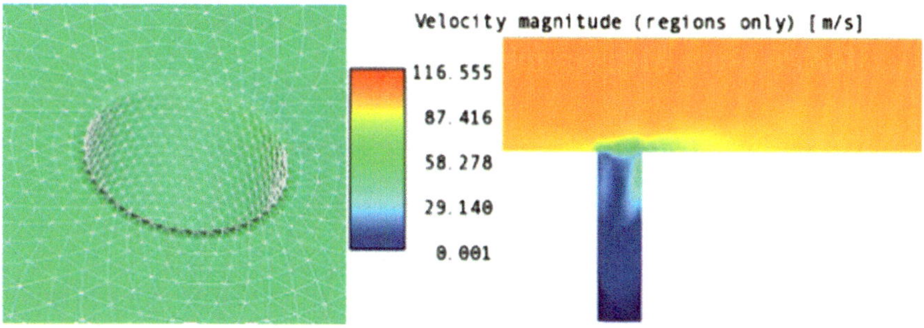

(a) Mesh division at the pressure orifice (b) Air velocity distribution near the pressure orifice

Fig. 7.6 Simulation of ideal hole pressure measurement with 90° vertical printing

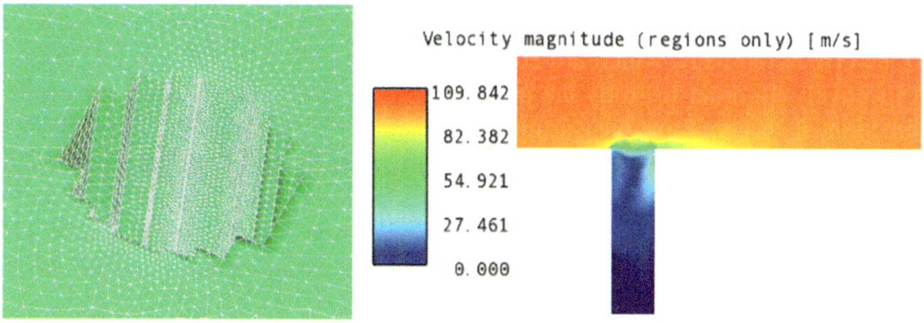

(a) Mesh division at the pressure orifice (b) Air velocity distribution near the pressure orifice

Fig. 7.7 Pressure measurement simulation of defective hole with 0° horizontal printing

Obviously, 3D printing is an accurate replica of the 3D theoretical model, and the vertical axis of the pressure measurement hole will not form excessive skew, but the step effect will change the surface quality of the hole. The actual measurement error brought by the 3D printing manufacturing, through the numerical simulation can only approximate the local schematic, in the future wind tunnel test shall analyze the test data to obtain sufficient experience and error correction.

Table 7.2 Flow rate and pressure gauges at the pressure measurement orifice and measurement points

Variables	90°Print hole P_t	0° print hole P_t	90°Print hole P_m	0° print hole P_m
Speed/m-s^S	50.05	51.77	0.0	0.0
Static pressure/Pa	131.98	112.57	411.52	351.91
Total pressure/Pa	2142.23	2054.85	411.53	351.92

7.3 Additive Manufacturing Processing of Models for Pressure Measurements

7.3.1 *Pressure Gauge Orifice Printing Experiment*

When manufacturing a pressure measurement model using 3D printing technology, the pressure transfer orifice is the key to printing. In engineering applications, a model component has anywhere from tens to hundreds of pressure transfer orifices. During the integrated manufacturing process of 3D printing, if one of the orifices fails to print or becomes clogged, the entire model will fail, so the orifice processing capability of 3D printing must be studied and an orifice structure suitable for 3D printing must be designed so that the pressure transfer model can be processed quickly and efficiently by 3D printing.

The orifice test model shown in Fig. 7.8 was divided into five groups according to the orifice radius R, which are 300 μm, 350 μm, 400 μm, 450 μm and 500 μm respectively. In order to study the printing angle α (the angle between the printing direction of the pressure measuring flow channel and the horizontal direction, as shown in Fig. 7.9) on the pressure orifice printing, five groups of models were printed at five angles of 0°, 10°, 20°, 30°, and 90°. In order to investigate the effect of runner length l on the pressure orifice printing, two groups of runner lengths (100 and 150 mm) were designed for comparison. The experimental parameters were designed as follows Table 7.3.

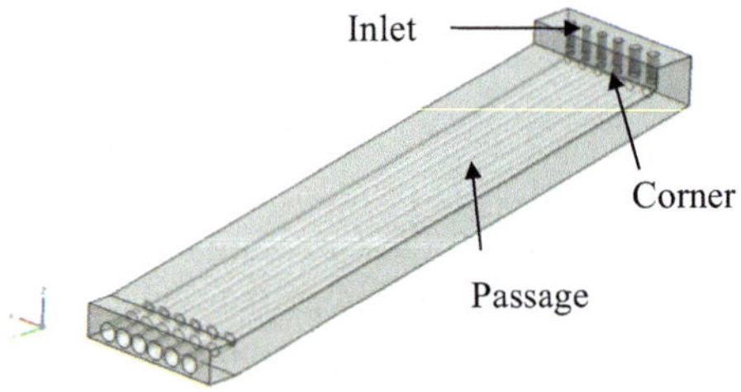

Fig. 7.8 Test model of pressure orifice

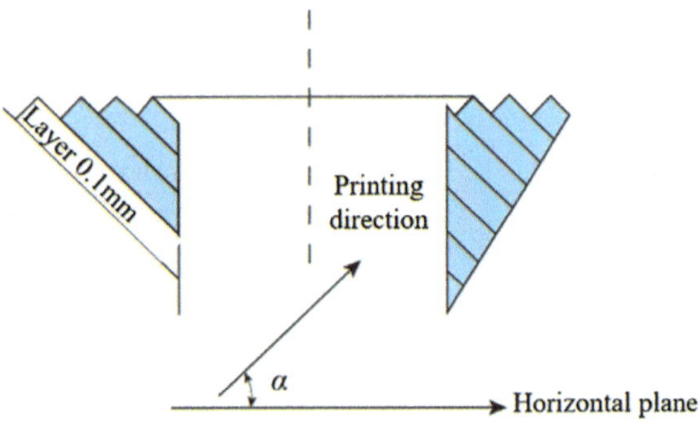

Fig. 7.9 Print direction diagram

Table 7.3 Parameter design table for the borehole test model

Test model group	Orifice radius/μm	Flow channel length l/mm	Print direction α/°
1	300	50/100/150	0/10/20/30/90
2	350	50/100/150	0/10/20/30/90
3	400	50/100/150	0/10/20/30/90
4	450	50/100/150	0/10/20/30/90
5	500	50/100/150	0/10/20/30/90

7.3.2 Analysis of Experimental Results of Orifice Printing

The light-curing rapid prototyping equipment used in this study is the SPS600B light-curing prototyping machine of Xi'an Jiaotong University. The model manufactured with this printer is shown in Fig. 7.10.

1) Conductable orifice study

After the model is manufactured, uncured liquid resin is adsorbed on the surface. If this resin cures, it will not only damage the surface finish of the part, but also cause dimensional and shape errors. Therefore, it is important to wait for a period of time after the pallets are raised before removing the part so that the liquid resin adsorbed on the surface of the model, especially in the pores, can flow back into the resin tank. In addition, the resin adsorbed inside the orifice must be cleaned with alcohol or acetone as soon as possible after the model is removed. In the pore channel conduction process, syringe injection and compressed air injection were used. The effect of pore diameter and printing direction on the conductivity of the pore channel was obtained from the printing results of groups 1 to 5 specimens as follows Table 7.4 The influence of aperture diameter and runner length on the conduction of the pressure orifice is shown in Table 7.4, and the influence of aperture diameter and runner length on the conduction of the pressure orifice is shown in Table 7.5 The effects of aperture diameter and flow path length on the conductivity of the pressure orifice are shown in Table 7.5. √ and × indicate conduction and non-conduction, respectively.

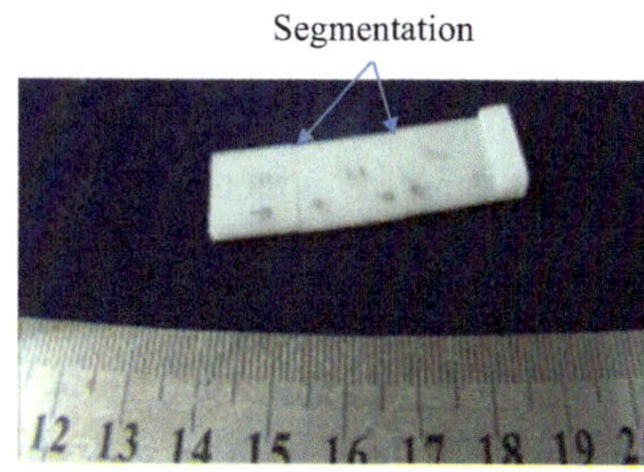

Fig. 7.10 Test model for pressure

Table 7.4 Relationship between orifice diameter, print direction and gauge orifice conduction or not (flow path length 50 mm)

Radius R/μm	Print direction α/°				
	0	10	20	30	90
300	×	×	×	×	×
350	×	×	√	√	√
400	×	√	√	√	√
450	√	√	√	√	√
500	√	√	√	√	√

Table 7.5 Table 2.3 Relationship between orifice diameter, flow path length and pressure measurement hole conduction or not (printing direction 90°)

Radius R/μm	Flow channel length l/mm		
	50	100	150
300	×	×	×
350	√	×	×
400	√	√	×
450	√	√	√
500	√	√	√

According to Tables 7.4 and 7.5 the results obtained, the following conclusions can be drawn:

(1) The orifice printed in the 0 degree direction is the most difficult to conduct, and 90 degrees is the easiest;
(2) The difficulty of conducting the orifice increases as the radius of the orifice decreases;
(3) The difficulty of conduction of the orifice increases with the increase of the length of the flow channel;
(4) When designing the orifice radius according to the length of the flow channel, the following principles should be followed: when the length of the flow channel is less than or equal to 50 mm, the minimum orifice radius that can be conducted is 350 μm; when the length of the flow channel is between 50 and 100 mm, the minimum orifice radius that can be conducted is 400 μm; when the length of the flow channel is greater than or equal to 100 mm, the minimum orifice radius that can be conducted is 450 μm.

2) Orifice quality analysis

Since traditional pressure measurement models use metal or plastic tubes with good internal surface quality, there is no clear quality requirement for the orifice for pressure transmission. However, the orifices manufactured by light-curing 3D printing

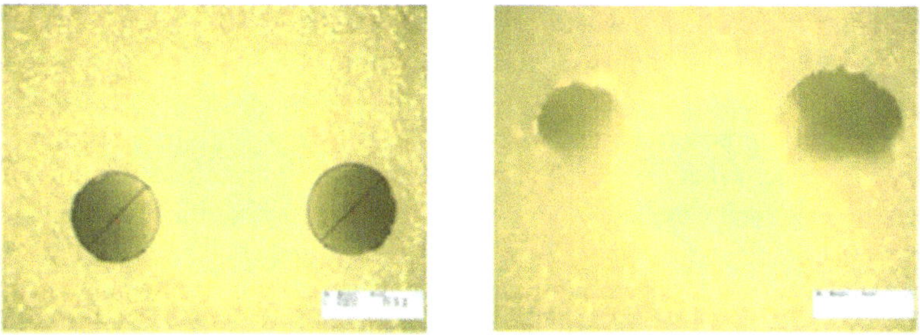

(a) Aperture R=350μm, printing direction 90° (b) Aperture R=350μm, printing direction 0°

Fig. 7.11 Borehole cross-section

have residual liquid resin inside them after printing and a small amount still adheres to the inner wall after cleaning, which has an impact on the quality of the orifices. In this section, the relationship between the printing direction and the quality of the aperture is investigated.

By Table 7.4, It can be seen that when the aperture diameter R is greater than 350 μm, the apertures made in any printing direction can be conducted, but the step effect of the light-curing printing process also makes the resin cleaned incompletely and attached to each step class to cause some impact on the quality of the apertures. Figure 7.11 the cross-sectional view of the pore channel for a model with a pore diameter R of 350 μm is shown in Fig. 7.11 at 0° and 90° printing angles, respectively. It can be seen that when the printing angle is 90 degrees, the aperture shape is close to circular and the aperture quality is good. When the printing angle is 0 degrees, the pore shape is elliptical and the pore quality is poor.

The printing direction of a porous part should be maximized in the direction of the hole axis. The finer the subdivision of the same surface, the better the quality of the curved surface. Therefore, when the layered thickness is a constant value, the printing direction should maximize the number of surface layups. In addition, the higher the height of the part, the longer the corresponding manufacturing time, the greater the laser power loss brought about, and the higher the manufacturing cost, so the test manufacturing direction is generally taken as 20º–30º for the test. When performing engineering applications, the part should be printed in a stack along the direction of the largest dimension due to the high requirements for surface quality.

7.3.3 Pressure Gauge Orifice Printing Accuracy Analysis

The dimensional accuracy of the models manufactured by the light-curing printer is determined by the printer. For miniature pressure measurement orifices, there is a certain deviation between the actual size of the model orifice and the design size. Therefore, an accuracy analysis of the orifice test model is necessary to objectively evaluate the dimensional deviation between the light-cured 3D printed model and the design value, and thus provide a compensation guideline for designing the size of the piezometric orifice.

The manufactured orifice test models were truncated into three sections along the length of the orifice, and each section was trimmed and cleaned to measure its orifice diameter and calculate its average value. The measurement equipment used was the VH-8000 microscope system from Keyence, Japan, which has a maximum magnification of 3000 and a minimum dimension of 50 nm, allowing for dimensioning and measurement, and direct output of digital images, as shown in Fig. 7.12. Table 7.6 shows the average values of the orifice radius for each group of models at a printing angle of 20 degrees and the average values of the standard deviation of each hole, which summarize the specific design compensation values. For example, if the runner length is 50 mm and the design orifice diameter is 450 μm, the average value of the measured orifice diameter is 411 μm and the compensation value is 44 μm, which means that the design value of the orifice diameter under this condition is equal to the expected value of the orifice diameter plus the compensation value.

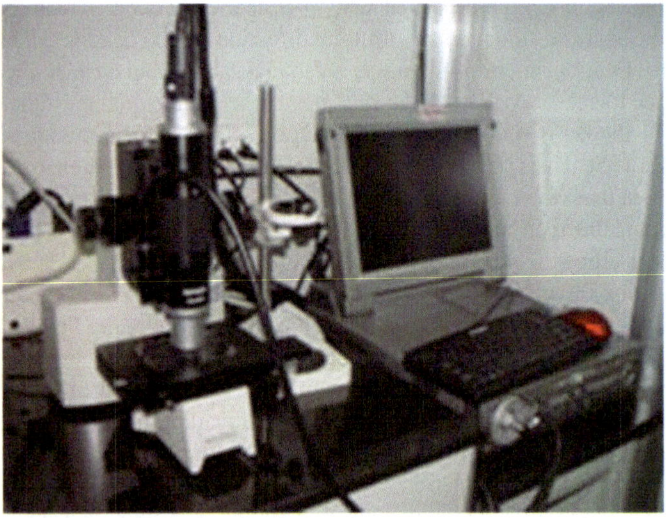

Fig. 7.12 VH-8000 microscope

Table 7.6 Pressure measurement orifice compensation table (20° printing direction)

Flow channel length l/mm	Pore size design value R/μm	Average value of pore size measurement/μm	Compensation value/μm
50	350	306	44
	400	358	42
	450	411	39
	500	465	35
100	400	348	52
	450	404	46
	500	459	41
150	450	391	59
	500	444	56

7.4 Connetctions of Models for Pressure Measurements

7.4.1 Study on the Connection Scheme of the Internal Pressure Measurement Orifice of the Model

In order to connect the orifices between the different components of the gauging model (fuselage, wing, front wing, rudder surface), the following two interfaces for the internal gauging orifices of the model were designed:

1) Metal tube—hose connection design

As shown in Fig. 7.13 in A, a 1.2 mm diameter hole is designed near the connection of the orifice to insert a metal tube with 1.2 mm outer diameter and 0.8 mm inner diameter, and in B, a 1.6 mm diameter hole is designed near the connection of the orifice to insert a plastic hose with 1.6 mm outer diameter and 1.2 mm inner diameter. The gap is filled with adhesive, and the metal tube in A and B is inserted into the plastic hose in B during assembly to complete the aperture connection. This kind of interface is for the smaller structure of flaps, ailerons, small rudder surfaces and other positions.

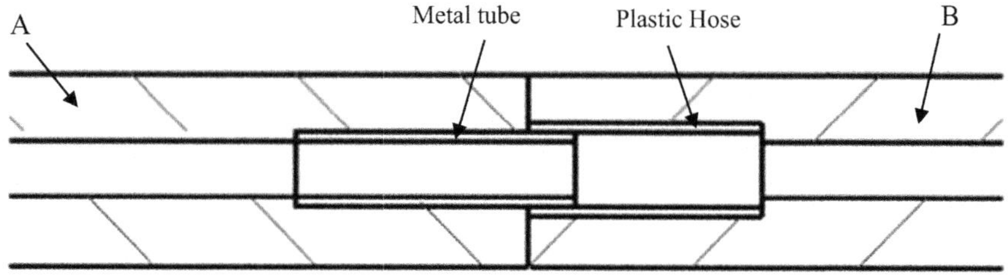

Fig. 7.13 Schematic diagram of metal tube—hose connection

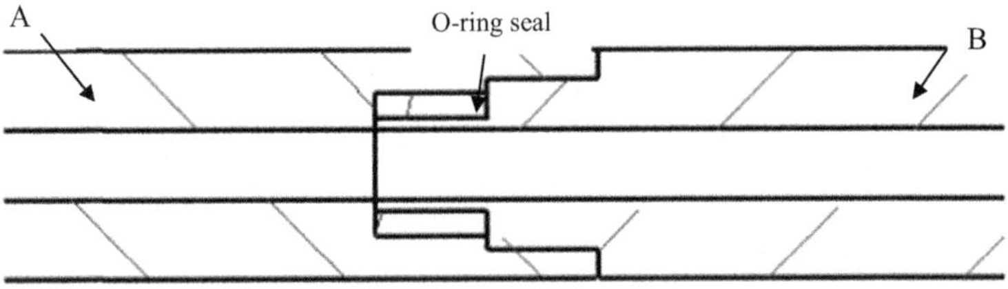

Fig. 7.14 Schematic diagram of O-ring connection

Table 7.7 Comparison of two internal orifice connection schemes

	Small structure connection, relative rotation	Large structure connection, fixing
Metal tube-hose	◆	
O-ring seal		◆

2) O-ring connection design

Part A and B of the model shown in Fig. 7.14 were assembled, and the orifice diameter is 0.8 mm. A groove was designed near the connection in part A, and a tab was designed near the connection in part B. An O-ring with an inner diameter of 2.0 mm and an outer diameter of 5.4 mm was installed in the radial direction. This type of interface is for larger structures such as fuselage, wing and other locations for connection.

3) Comparison of the two solutions

A comparision of the above two internal orifice connection solutions is shown in Table 7.7 based on the structural characteristics.

7.4.2 *Study on the Connection Scheme of the Internal Pressure Measurement Orifice of the Model*

To perform the pressure measurement wind tunnel experiments, it is necessary to connect the pressure measurement orifice in the model to the external pressure measurement system, so two interface solutions are designed: single-hole connection and modular connection.

1. Single-hole interface design

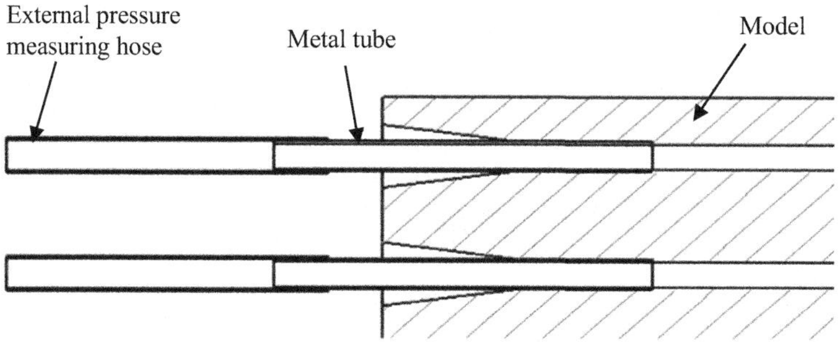

Fig. 7.15 Schematic diagram of metal tube interface

The single-hole interface options are divided into the following three types:

1) Metal pipe interface. As shown in Fig. 7.15 the metal pipe is sealed with adhesive at the gap between the metal pipe and the model, and the metal pipe is inserted into the external pressure measurement hose for connection.
2) Hose interface. As shown in Fig. 7.16, a conical hole is designed at the outlet of the model orifice, and the pressure measurement orifice in the model is extended outward, and the 1.6mm outer diameter and 1.2mm inner diameter pressure measurement hose is inserted into the hole to wrap the outward extension of the pressure measurement pipe, and the gap is filled and sealed with adhesive.
3) Resin interface. As Fig. 7.17 shown, a section of pipe extending outward is designed at the exit of the model orifice and printed together with the model. The pipe is divided into two sections, the section near the model is a transition section with a 5-degree inclination angle, and then the interface section with an outer diameter of 1.2 mm and an inner diameter of 0.8 mm. When connecting, a 1.6 mm outer diameter and 1.2mm inner diameter pressure measurement hose is placed on the interface section to complete the connection with the external pressure measurement system.

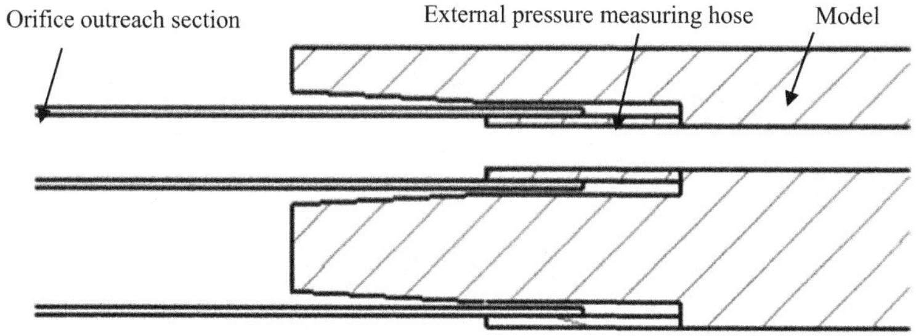

Fig. 7.16 Schematic diagram of the hose connection

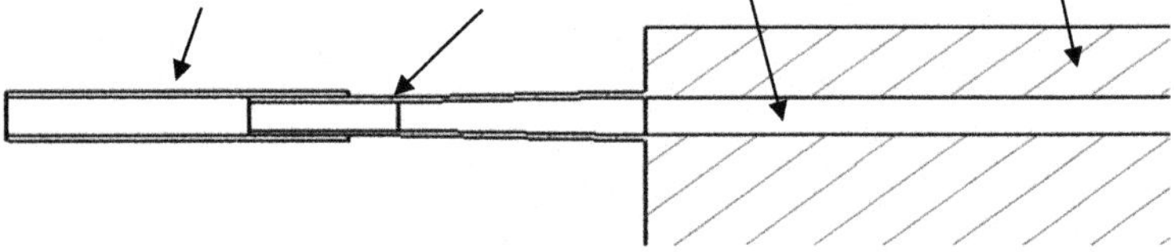

Fig. 7.17 Schematic diagram of resin interface

2. Modular interface design

Due to the large number of pressure measurement orifices in the wind tunnel test model, the following two modular interface solutions were designed to connect multiple orifices at once in order to improve the connection efficiency.

1) Modularization scheme I. As shown in Fig. 7.18, a conical hole is designed at the outlet of the model orifice, and the position and orientation of the pressure measuring hose is fixed by a light-curing formed cover plate with holes on the outside of the model, and the cover plate is docked with the model when connecting, and the pressure measuring hose is inserted into the conical hole of the model, and the cover plate is filled with adhesive between the hose and the cover plate.
2) Modularization scheme II. As shown in Fig. 7.19, the resin cover plate is used to fix the position and orientation of the pressure measurement hose, and a 1.2 mm diameter hole is designed at the outlet of the model orifice.

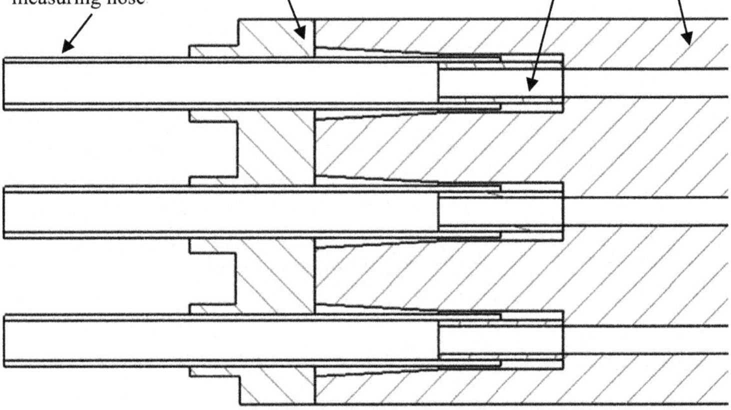

Fig. 7.18 Schematic diagram of modular connection scheme I

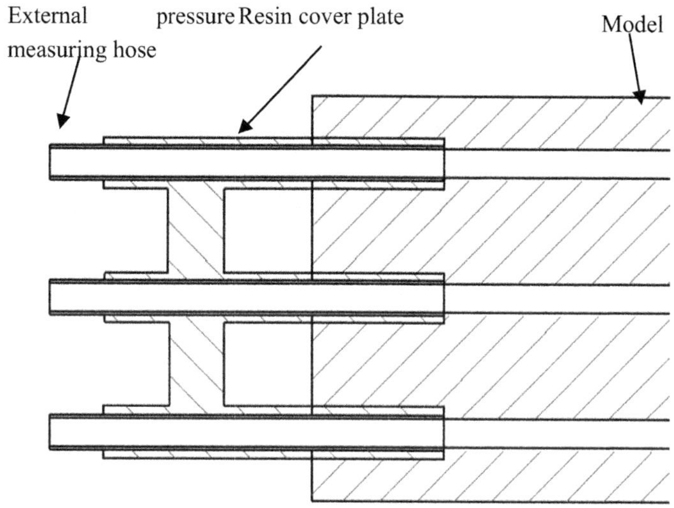

Fig. 7.19 Schematic diagram of modular connection scheme II

3. Comparison of five interface solutions

The above five external interface solutions are compared in five aspects: simple structure, solid bonding, detachable, easy cleaning and connection efficiency, and their characteristics are summarized as follows Table 7.8.

Table 7.8 Comparison of the external interface characteristics of the pressure measurement orifice

Interface Solutions	Metal tube interface	Hose connection	Resin interface	Modularity I	Modularization II
Simple structure installation	◆	◆	◆		
Firm bonding	◆				◆
Detachable		◆	◆	◆	
Facilitates orifice cleaning	◆	◆			◆
Connection efficiency				◆	◆

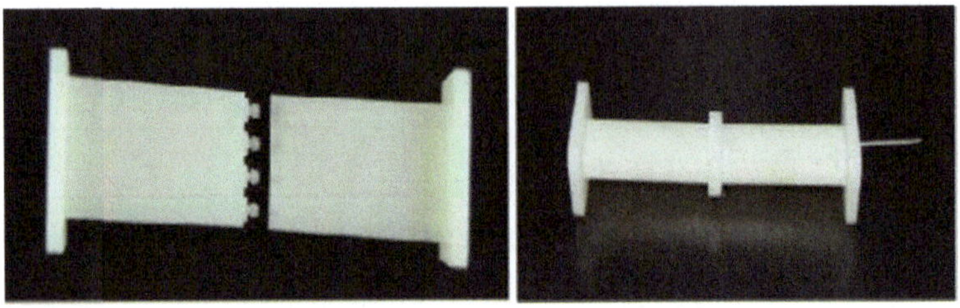

(a) O-ring connection (b) Metal tube-hose connection

Fig. 7.20 Internal orifice connection model

7.4.3 Performance Study of Pressure Measurement Orifices and Interfaces

Based on the study of the connection scheme of the pressure measurement orifice, models of the pressure measurement orifice with various interfaces were designed and manufactured with a length of 150 mm and an orifice radius of 0.4 mm, as Figs. 7.20, 7.21, and 7.22

1) Gas tightness test verification

According to "GJB 180A-2006 A Low-speed Wind Tunnel Aircraft Model Design Specification", when there is a pressure difference of not less than 3200 Pa between the inside and outside of the orifice, and the loss of ΔP between the inside and outside of the orifice within 60 s is less than 8 Pa, the requirement of airtightness of the pressure measurement orifice is met. Therefore, the schematic diagram of the experimental equipment designed to verify the air tightness of the pressure measuring orifice and the interface is as follows Fig. 7.23.

The experimental steps are as follows:

Turn on the high pressure gas source and set the pressure value to 105,000 Pa;
Opening valve B, closing valve A and blocking the test model orifice outlet;
Open valve A. After 60 s, read the pressure gauge reading and calculate the difference between it and the original pressure;

Repeat the above steps 3 times and calculate the average value as Table 7.9.

The experimental results show that the gas tightness of the model of the pressure orifice with three kinds of interfaces meets the standard of the wind tunnel experiments of pressure measurement.

2) Aeration experimental verification

According to "GJB 180A-2006 A Low-speed Wind Tunnel Aircraft Model Design Specification", when a jump-step pressure difference of not less than 3200 Pa occurs between the input and output of the pressure measurement pipeline, the pressure difference between the two ends of the orifice is less than 8 Pa within 2 s, which

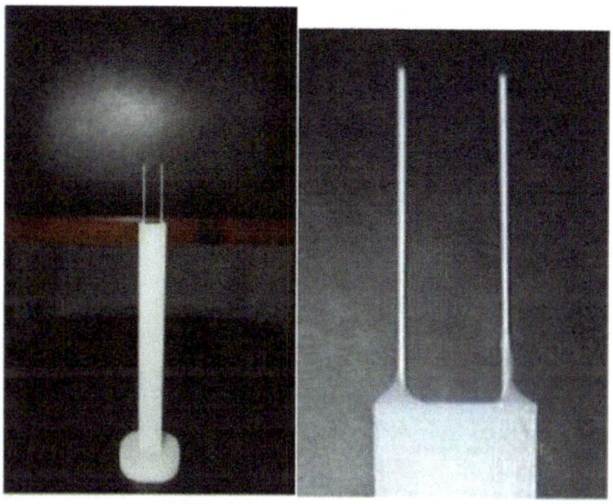

(a) Metal pipe interface

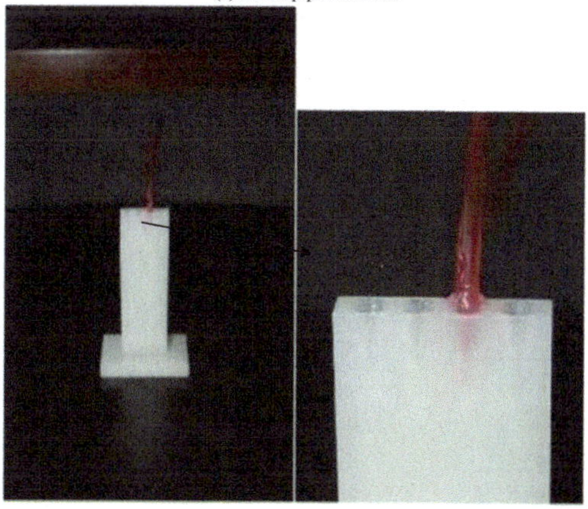

(b) Hose connection

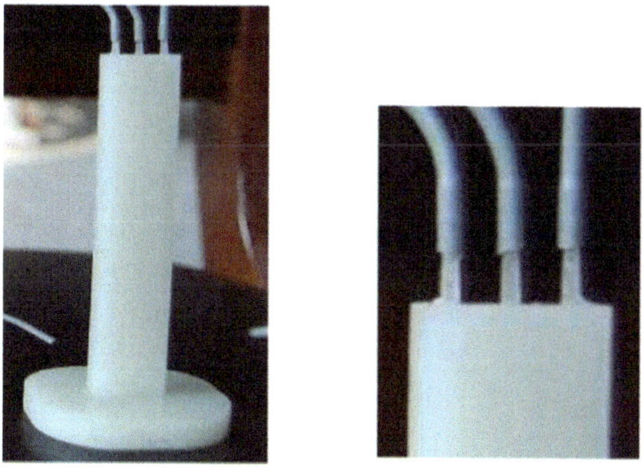

(c) Resin interface

Fig. 7.21 Single-hole interface model

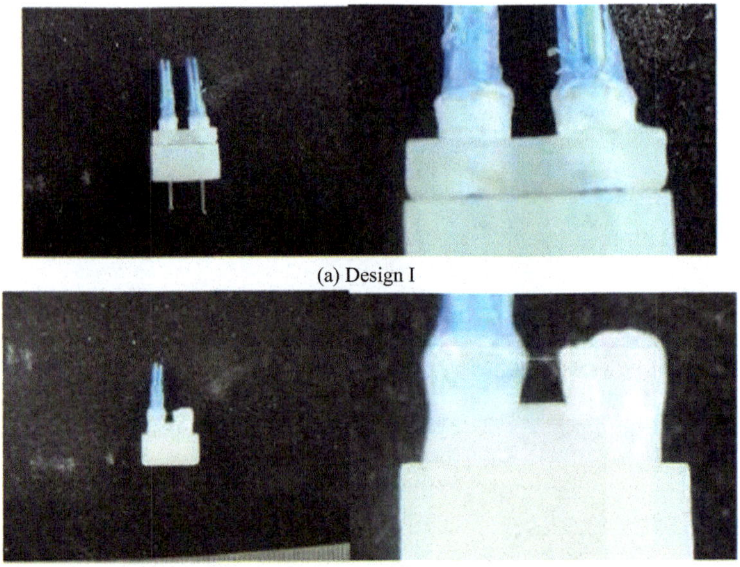

(a) Design I

(b) Deisgn II

Fig. 7.22 Modular interface model

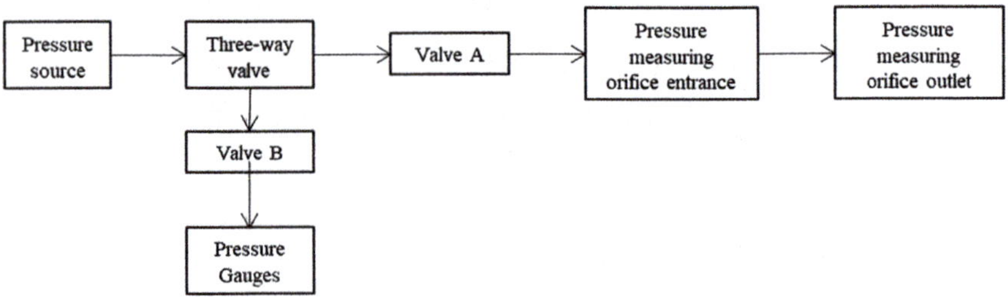

Fig. 7.23 Schematic diagram of the airtightness experimental setup

Table 7.9 Airtightness test results

Interface method	Pressure/Pa	ΔP/Pa
Metal tube interface	104,995	−5
Hose connection	104,994	−6
Resin interface	104,998	−2

meets the requirement of venting of the pressure measurement orifice. Therefore, the schematic diagram of the experimental equipment designed to verify the venting of the pressure measuring orifice and interface is as follows Fig. 7.24.

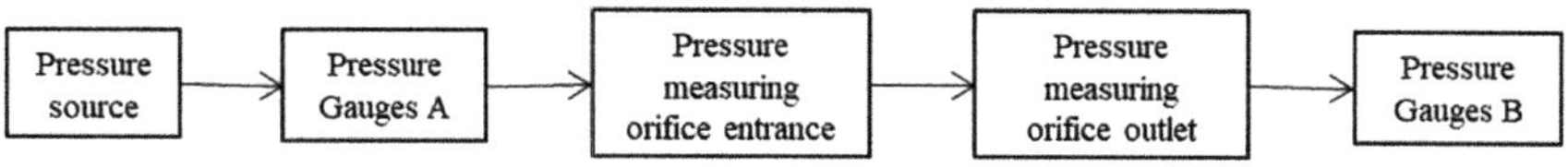

Fig. 7.24 Schematic diagram of the aeration experimental setup

Table 7.10 Aeration experiment results

Interface method	Pressure gauge A/Pa	Pressure gauge B/Pa	$\lvert\Delta P\rvert$/Pa
Metal tube interface	104,995	104,994	1
Hose connection	104,994	104,990	4
Resin interface	104,998	104,995	3

The experimental steps are as follows:

Turn on the high pressure source, set the pressure value to 100000Pa, at this time the pressure gauge A, B readings are 10000Pa;
Regulating the pressure value to 105,000 Pa at a certain point t;
Read the readings of pressure gauges A and B respectively at t + 2 s and calculate the difference between them;
Repeat the above steps 3 times and calculate the average value as shown in Table 7.10.

The experimental results show that the venting of the model with three kinds of interfaces meets the standard of the pressure measurement wind tunnel experiments.

Chapter 8
Additive Manufacturing of Wind Tunnel Test Models for Static Aeroelastic Properties

8.1 Overview

The model used in the static aeroelastic wind tunnel test is the static aeroelastic model [88]. The ideal static aeroelastic model has the same aerodynamic shape and similar structural parameters (mainly structural stiffness) as the prototype aircraft in terms of scaling. Currently, static aeroelastic models often use beam frame—frame section structure. The main components of the beam frame structure are single or multiple beams, located inside the model (fuselage and wings), generally made of metal, which simulate the stiffness of the model and provide strength to the overall model; the frame segment structure is mounted on the beam frame, which forms the aerodynamic shape of the model and transmits the aerodynamic force to the beam frame. For design and processing convenience, the stiffness of the model is entirely provided by the beam frame, and the complete aerodynamic shape is divided into disconnected frame segments (e.g., Fig. 8.1). The complete aerodynamic shape is divided into disconnected frame segments (shown in Fig. 8.1) and mounted on the beam frame as "point connections" to reduce the influence of the shape stiffness on the design model stiffness. During the blow test, the frame segments were filled with low modulus materials such as foam and elastomer to maintain the complete aerodynamic shape. In the low-speed stage, the model designed and processed by the above scheme can provide enough accurate test data; however, in the high-speed test, the low-modulus material between the frame segments will be deformed greatly, and the integrity of the aerodynamic shape is difficult to be guaranteed, which increases the uncertainty of the test data.

After more than two decades of development, 3D printing has been applied to the manufacturing of wind tunnel test models for a wide range of aircraft, speed ranges, and test types, and its feasibility and advantages have been verified in large numbers. However, there are relatively few reports on the application of elastic models, and there are many reasons for this situation. First, because the elastic model test is an "advanced test project", its implementation is in the final stage of aircraft design,

W. Zhu and D. Li, *Models for Wind Tunnel Tests Based on Additive Manufacturing Technology*, Additive Manufacturing Technology,
https://doi.org/10.1007/978-981-99-5877-1_8

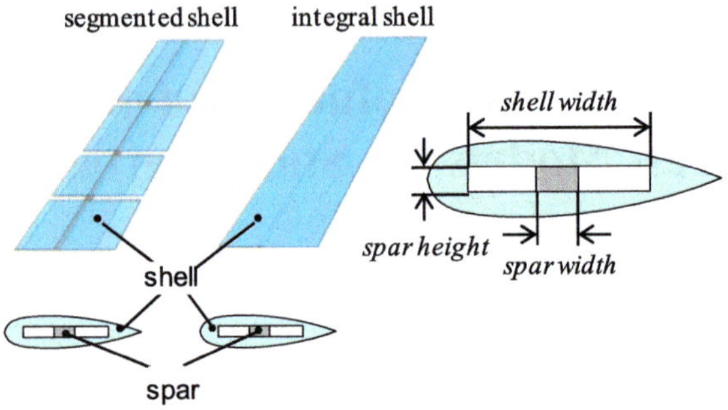

Fig. 8.1 Frame section housing and integral housing for the pneumatic model [89]

which requires high confidentiality and is not easy to be reported publicly; second, the elastic model has more design constraints and specific similar requirements for structural parameters, and its design and manufacturing process is tedious for the feasibility verification project. This research object is too complex, so it is generally not selected as a research topic; furthermore, elastic models are sensitive to the mechanical response of materials, and the replacement of manufacturing materials and manufacturing processes has a greater impact. Compared with general rigid models, the selection of 3D printing technology to process elastic models, engineers have to take greater risks, which also limits the adoption of this type of models to new technologies. In summary, the need for confidentiality, design difficulty, and implementation risk have resulted in fewer reports of research on the adoption of 3D printing technology for elastic models. The limited reports focus on the use of 3D printing to replace non-metallic material components [49]. The model simulates the dynamic characteristics of the wing section of a prototype aircraft with a beam-frame segment structure, in which the aerodynamic shape is divided into disconnected segments. Tests have shown that the plastic frame segments machined with SL can withstand transonic loads and satisfactory data can be obtained.

This chapter proposes a structural solution and implementation method for a wind tunnel test model with an integrated aerodynamic shape. The key of this new idea is that the aerodynamic shape is an integrated structure that contributes to the overall stiffness of the model and is machined in one piece by 3D printing after the design is completed; the beam frame structure can be simplified. The idea can solve the influence of the deformation of the segmented frame section on the aerodynamic shape of the model on the one hand, which is expected to improve the geometric similarity of the model and thus ensure the accuracy of the wind tunnel test; on the other hand, it simplifies the model design and manufacturing process and helps to improve the efficiency of the wind tunnel test.

The structure of the model is shown in Fig. 8.2. The structure of the model is shown in Fig. 8.2, which consists of two major parts: the metal skeleton and the plastic shell. The metal skeleton is located inside the model and divided into two major parts, fuselage and wing, which provide the main strength and stiffness and are mounted on the balance located at the tail of the model through the inner cavity of the fuselage; the plastic shell is divided into two major parts, fuselage and wing, which

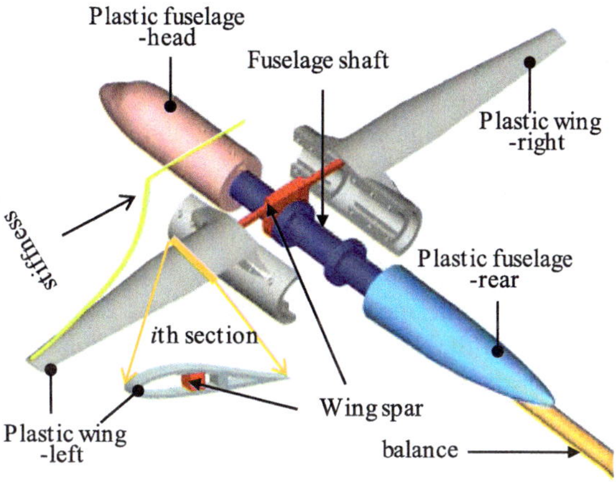

Fig. 8.2 Structural scheme of the composite aircraft model

are mounted on the corresponding metal skeleton and provide the full aerodynamic shape of the model. As shown in Fig. 8.2, both the left and right wing shells are integrated monolithic structures, which are machined by 3D printing (SL process) in one go. Because of their large stiffness, the impact on the overall wing stiffness cannot be neglected, so the stiffness of the plastic shells needs to be accounted for in the stiffness design. To maintain the continuity of the aerodynamic shape as much as possible, the fuselage part connected to the wing is also integrated into the wing, designed and machined in one piece, thanks to the powerful manufacturing capability of 3D printing technology for complex structures inside and outside. Other plastic shells include the front and rear sections of the fuselage—the nose and fuselage—and since the fuselage stiffness did not need to be simulated in this project, the design of this section was made simple, requiring only the maintenance of the correct aerodynamic shape and a reliable fit to the rest of the model.

8.2 Composite Model Design Method

8.2.1 Design Process

From the above, it is clear that the key to the design of this model is the wing part—the load-bearing metal beam frame and the integrated plastic shell. Since the wing of the prototype with a large span is long and narrow, and the variability along the chordal direction within the airfoil is small, its structural parameters can be simplified to a single beam, so the metal beam frame of this model adopts a single beam structure.

As shown in Fig. 8.2 the left and right beams are joined as one at the wing root and fixed to the fuselage skeleton by the assembly, and only the part from the wing tip to the wing root needs to be designed for the model design. The plastic shell is installed on the outside of the beam, and the left and right parts are installed in the middle of the fuselage in a docking position. Because of its integrated structure, it has a large stiffness and contributes to the wing stiffness, and its influence needs to be considered in the model design phase.

The design flow of the static aeroelastic model is shown in Fig. 8.3. From the prototype to the model goes through two stages of similar design and structural design, and the design objects include the structural parameters and aerodynamic shape of the model and the aerodynamic parameters of the blowing wind. In order to verify the model design accuracy, a series of validation tests were conducted in this project, including ground deformation tests for the structural parameters of the model and numerical simulations and wind tunnel tests for the overall performance of the model.

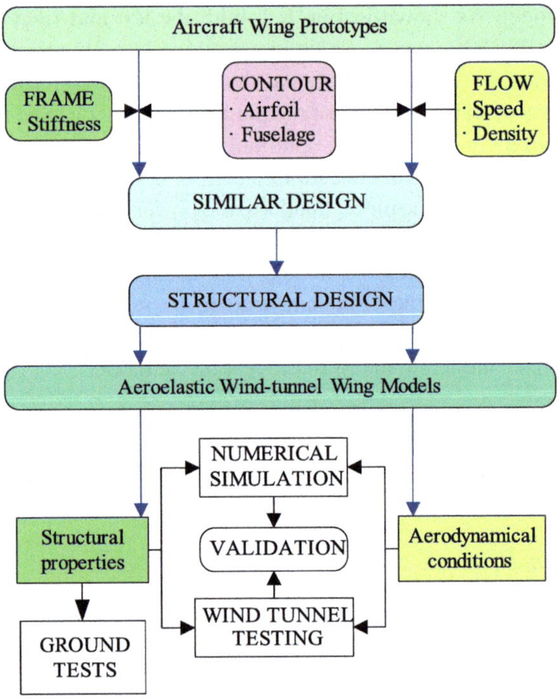

Fig. 8.3 Design flow of the pneumatic elastic model

8.2.2 Scaled Down Design

The task of similarity design is to determine the scaling factors of the model. For static aeroelastic models, the scaling factors to be determined are divided into two categories: dimensional scaling factors and aerodynamic scaling factors: the former includes form geometry and structural parameters; the latter covers Mach number, dynamic pressure, and density. Form geometry scaling factor K_l is determined according to the prototype size, wind tunnel test section size and sky level parameters; the Mach number *Ma* of the model should be consistent with the prototype, i.e., the scaling factor is 1; the dynamic compression ratio coefficient K_q and density scaling factor K_ρ. For the static aeroelastic model, the scaling factor of structural parameters refers to the ratio of structural stiffness between the prototype and the model, and the large span ratio wing mainly includes the horizontal bending stiffness scaling factor of the wing section K_{EI} and torsional stiffness scaling factor K_{GJ} For the large span ratio wing, the horizontal bending stiffness scaling factor and the torsional stiffness scaling factor of the wing section are mainly included.

8.2.3 Stiffness Design

In this section, based on the model dimensions and structural stiffness distribution data obtained above, we detail the selection of a suitable structure and the determination of the corresponding dimensions to achieve the obtained stiffness distribution.

1) Cross-sectional structure

As mentioned above, the single wing consists of a single metal beam and a plastic shell. Due to the integrated structural characteristics of the plastic shell, its stiffness is not negligible and its contribution to the overall stiffness of the wing needs to be considered in the design. The internal structural solution of the wing selected for this project is shown in Fig. 8.4.

For the convenience of design and processing, the metal beam is selected with rectangular cross-section, and its shear center coincides with the shear center of the prototype wing section. For the integrated plastic shell, the shape geometry is obtained from the prototype according to the dimensional scaling factor, and the internal cavity is cavity to set the mounting surface of the metal beam. To improve the stiffness and strength of the shell, the vertical web is designed at the trailing edge position to divide the cavity into two independent closed forms.

2) Cross-sectional dimensions

As mentioned above, the horizontal bending stiffness *EI* (EI_{xx}) and torsional stiffness *GJ* (GJ_{zz}) are distributed continuously along the spreading direction, and ideally, the structure to be designed should have exactly the same distribution continuously.

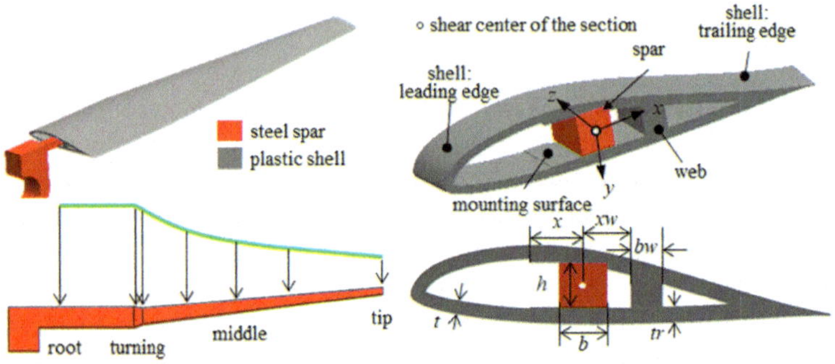

Fig. 8.4 Stiffness design of the wing section

In fact, for reasons of design complexity and processing economy, in engineering practice, a number of key sections are generally selected along the spreading direction to make the stiffness distribution reach the target value, and the structural dimensions of the key sections are approximated by the interpolation method. This method can ensure that the stiffness distribution of the critical section has high accuracy and the error of the interpolation part can be better controlled, and also ensure that the designed structure has good processability. As shown in Fig. 8.4 seven critical sections were selected along the wing span: one at the wing root and one at the wing tip, two at the sharp change in stiffness and three at the intermediate transition section.

In order to complete the design of the model wing structure, the dimensions of the above structure need to be determined. In this project, the structural dimensions of the selected cross-section, including the position of the cross-sectional torsional center and the structural dimensions of the beam and the shell, are designed using an optimal design method. The position of the torsional center of the cross-section of the model includes the respective torsional centers of the metal beam and the plastic shell, which coincide with the torsional center of the prototype after scaling. The structural dimensions of the beam and shell are shown in Fig. 8.4, driven by the model geometric parameters section moment of inertia I and section torsion constant J. *The* horizontal bending stiffness EI and torsional stiffness GJ *of* each key section have been obtained in the scaled-down design stage. The optimized configuration is given by the following equation.

$$\text{minimize} f(\boldsymbol{x}) \tag{8.1}$$

The expression of the objective function is shown below as the error-weighted sum of three parts of the optimization objective: horizontal bending stiffness EI, torsional stiffness GJ and torsional center position SC. α_{EI}, α_{GJ} and α_{SC} are the weighted amounts of each objective quantity, which are taken from the (0,1) open interval, respectively, and the sum is 1.

$$f(\boldsymbol{x}) = \alpha_{EI}\sum_{i}^{n}|\Delta EI_i| + \alpha_{GJ}\sum_{i}^{n}|\Delta GJ_i| + \alpha_{SC}\sum_{i}^{n}|\Delta SC_i| \tag{8.2}$$

where i is the serial number of the critical section and n is its total number, n = 7 in this project; $|\Delta EI_i|/|\Delta GJ_i|$ is the absolute value of the error between the sum of horizontal bending stiffness/torsional stiffness of the metal beam and plastic shell of the *ith critical* section and the target value, and its expression is as follows:

$$|\Delta EI_i| = \left|\left(EI_{\text{spar}} + EI_{\text{shell}} - EI_{\text{target}}\right)_i\right| \tag{8.3}$$

$$|\Delta EI_i| = \left|\left(GJ_{\text{spar}} + GJ_{\text{shell}} - GJ_{\text{target}}\right)_i\right| \tag{8.4}$$

Similarly, an expression for the error in the torsional center of the *ith* critical section can be obtained as

$$|\Delta SC_i| = \left|\left(SC_{\text{spar}} - SC_{\text{target}}\right)_i\right| + \left|\left(SC_{\text{shell}} - SC_{\text{target}}\right)_i\right| \tag{8.5}$$

where *SC* includes the coordinate values in the x and y directions, as Fig. 8.84 is shown.

The optimization variables include the amount of metal and the geometric parameters of the plastic housing, expressed as follows:

$$\boldsymbol{x} = \left(\boldsymbol{x}_{\text{spar}}, \boldsymbol{x}_{\text{shell}}\right) \tag{8.6}$$

where the structural dimensions of the metal beam include width *s_w* and height *s_ h*, as follows:

$$\boldsymbol{x}_{\text{spar}} = (s\_w, s\_h) \tag{8.7}$$

The structural dimensions of the plastic housing include the mounting surface positioning coordinates *m_x* and height *m_h*, the web positioning coordinates *w_x* and thickness *w_t*, and the thickness of the leading and trailing edges of the housing *l_t* and *t_t*. Among them, the mounting surface height *m_h* is the same as the metal beam height *s_h*, and this variable can be incorporated into the beam. Therefore, the optimized variables for the plastic housing are as follows.

$$\boldsymbol{x}_{\text{shell}} = (m\_x, w_x, w_t, l\_t, t\_t) \tag{8.8}$$

In order to obtain a reasonable solution, it is necessary to set a certain value interval for the optimization variables.

8.3 Case Studies

8.3.1 *Model Design*

The static aeroelastic model in this study is a 1/35-scale wing-body model of an airliner to simulate the stiffness distribution of the wing section of the prototype aircraft, to test the component deformation and aerodynamic response of the model under specific conditions, to determine the divergence conditions, and to provide basic data for aircraft prototype design. The structure of the model is shown in Fig. 8.2.

1) Scaled down design

The shape of the model can be obtained by simply scaling down the geometry of the prototype according to the dimensional reduction factor K_l The Mach number *Ma* and dynamic pressure parameter q in the wind tunnel flow field can be obtained by adopting the values used in the scaling design; as for the internal structure, a suitable structure can be designed according to the stiffness distribution, and the stiffness distribution is obtained from the prototype stiffness distribution and the stiffness scaling factor K_{EI}/K_{GJ}. The distribution of the horizontal bending stiffness *EI* along the spreading direction is shown in Fig. 8.2, and the torsional stiffness distribution *GJ is* calculated in a similar way. The scaling factors used in this project are shown in Table 8.1.

2) Stiffness design

In this section, based on the model dimensions and structural stiffness distribution data obtained above, the selection of a suitable structure and the determination of the corresponding dimensions to achieve the obtained stiffness distribution are detailed. The material property parameters used are as follows Table 8.2.

The internal structure of the wing chosen for this project is shown in Fig. 8.4. For design and processing convenience, a rectangular cross-section was chosen for the metal beam, whose shear center coincides with the shear center of the prototype wing section. For the integrated plastic shell, the shape geometry is obtained from the prototype according to the dimensional scaling factor, and the internal cavity is

Table 8.1 Similarity criteria satisfied by the model design

Similarity criteria		Expressions	Similarity ratio[a]
Basic guidelines	Mach number	K_{Ma}	1
	Dynamic pressure	K_q	1.725
	Size	K_l	1/35
Guidelines for derivation	Stiffness	$K_{EI}/K_{GJ} = K_q K_l^4$	1.150E-6
	Load	$K_F = K_q K_l^2$	1.408E-3

[a]The similarity ratio is model-prototype

Table 8.2 Material performance parameters

Parameters	Photosensitive resin SOMOS14120	40Cr steel
Density, ρ /kg-m^{-3}	1120	7780
Young's modulus, *E*/GPa	2.46	211
Shear modulus, *G*/GPa	0.89	77
Poisson's ratio, ν	0.38	0.30
Tensile strength, σ /MPa	46.0	980

a cavity with the mounting surface of the metal beam set. To improve the stiffness and strength of the shell, a vertical web is designed at the trailing edge position to divide the cavity into two independent closed forms.

As mentioned above, the horizontal bending stiffness *EI* (EI_{xx}) and the torsional stiffness *GJ* (GJ_{zz}) are distributed continuously along the spreading direction. As shown in Fig. 8.4, seven critical sections were selected along the wing span: one at the wing root and one at the wing tip, two at the sharp change in stiffness and three at the intermediate transition section. The horizontal bending stiffness *EI* and torsional stiffness *GJ of* each critical section have been obtained in the scaling design stage Table 8.2. The material modulus of elasticity *E* and shear modulus *G are given*, according to which the required section geometry parameters can be obtained.

The result of the optimization calculation is shown in Fig. 8.5. Figure 8.5a gives the error of stiffness of each section, the error of horizontal bending stiffness is less than 5%, and the error of torsional stiffness is less than 2%, which is a small error level. From Fig. 8.5b, it can be seen that the metal skeleton stiffness is a proportion of the overall wing stiffness, and at the wing root, the plastic shell contributes a larger bending stiffness (about 15%) and torsional stiffness (about 25%).

3) Detailed design

The connection between the metal beam and the plastic shell needs to be designed to ensure a reliable and stable connection while minimizing the impact on the wing stiffness and taking into account the good machinability of the structure. In this book,

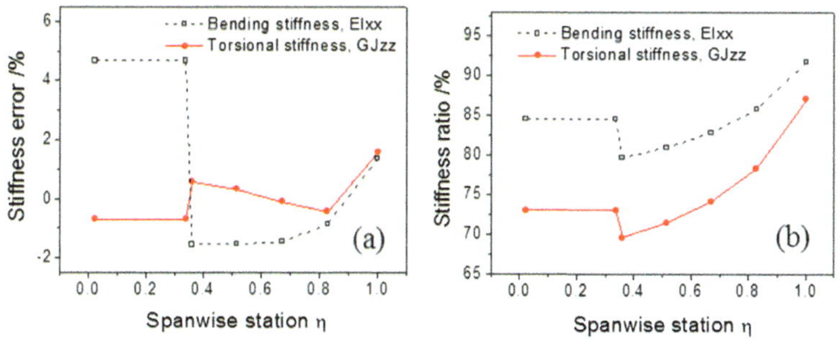

Fig. 8.5 Stiffness design results

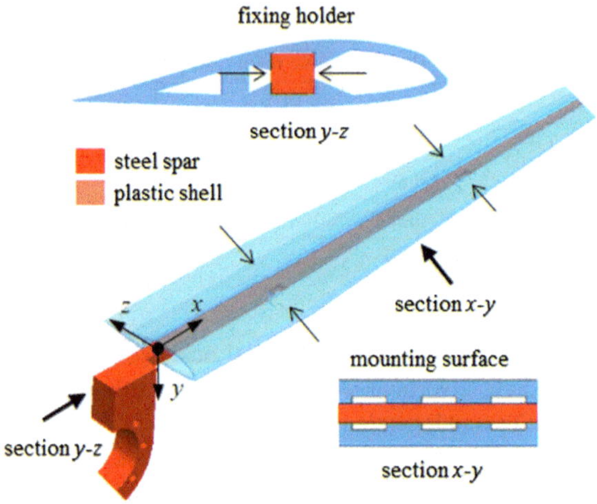

Fig. 8.6 Detailing of the model

the connection is realized by designing the wing rib fixed section and step mounting surface, and the structure is schematically shown in Fig. 8.6.

Two cross-sectional design beam ribs are selected in the middle section of the plastic housing for mounting metal beams. As shown in Fig. 8.6, four ribs protrude inward from the skin to fix the metal volume, and their dimensions are the same as those of the metal beam of the corresponding section to ensure the positioning accuracy. This wing rib fixing section ensures a precise and solid connection between the metal beam and the plastic shell, and achieves the transfer of bending and torsional forces between them. Along the wing span direction, the upper and lower parts of the shell and the beam are face fits, which increases the difficulty of machining and assembly. In this book, by setting grooves on the mating surface of the inner surface of the shell, a stepped mating surface is formed to reduce the mating area, and sufficient connection strength is ensured by designing the number and size of the grooves.

8.3.2 Strength Simulation Calibration

The model design must be strength calibrated, especially because the tensile strength of resin material is much smaller than that of metal material, and it is easy to deform or even fracture under large load conditions. In this book, CFD-CSD is used for strength check, and CFD is used to calculate the flow field for severe working conditions; the flow field load is loaded onto the corresponding part of the model, and CSD is used to calculate the static response.

Table 8.3 Parameters for pneumatic analysis

Parameters	Value
Mach number, *Ma*	0.4
Angle of attack, α /°	4.4
Incoming flow pressure,*P* /Pa	70,000
Reference temperature, *T*/K	265.6

1) Pneumatic analysis

The blowing test conditions to be carried out in this project are Mach number Ma from 0.4 to 0.65, angle of attack α from −4.0° to + 4.4° (angle of attack range varies depending on Mach number), because lift is the main load on the wing, and the angle of attack has a large impact on it, so the angle of attack with the maximum angle of attack is chosen as the calculation condition for the strength check, specifically Ma = 0.4, α = 4.4. Other aerodynamic parameters are shown in Table 8.3.

The computational grid used for CFD calculations is shown in Fig. 8.7. Due to the symmetry of the model, a symmetry plane was set along the longitudinal plane of the aircraft fuselage to reduce the computational effort. The computational mesh is a tetrahedral unstructured mesh with a total number in the order of 1 million.

The flow field model is given in Fig. 8.8, which was built with reference to the wind tunnel used, and the model was installed in the middle; one side of the flow field is the air inlet and three sides are the air outlet.

2) Strength calculation

The computational model is simplified into three parts: plastic shell, metal wing beam and metal fuselage skeleton, except for the tangential slip connection (No separation, ANSYS® Workbench®) between the wing shell and beam, all other mating surfaces are fixed connections (Bonded, ANSYS® Workbench®), and fixed constraints and Y-directional displacement constraints are set on the model's balance mounting The

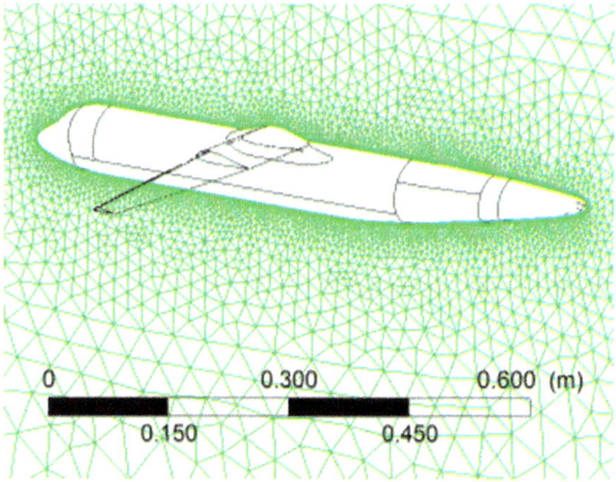

Fig. 8.7 Mesh used for CFD calculations

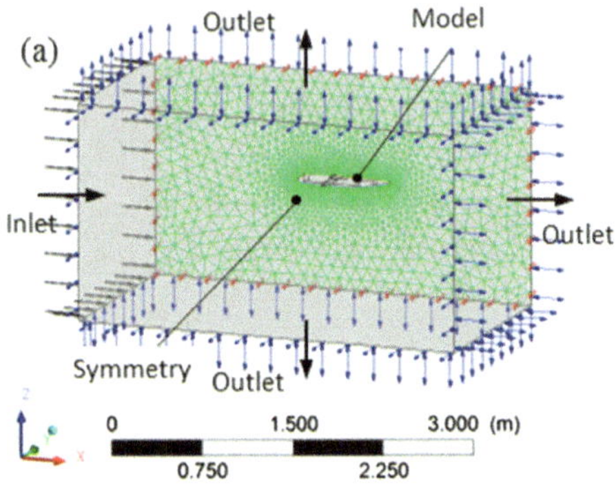

Fig. 8.8 Flow field used for CFD calculations

model mesh is a hexahedral and tetrahedral structure mesh with a total number of cells in the order of 100,000. The above fluid results are loaded onto the surface of the model, and the obtained deformation and stress results are shown in Fig. 8.9. The wing of the hydro-elastic model is flexible and has the same shape as Fig. 8.9a shows a large deformation displacement; the maximum stresses are all on the wing, and the maximum equivalent stresses of the metal beam are located at the wing root, while the maximum stresses of the plastic shell occur at the wing tip, so the skin is thin (~0.5 mm) there. It should be noted that the maximum stresses of both metal and plastic of the model are smaller than the required stresses and can be safely used for wind tunnel tests.

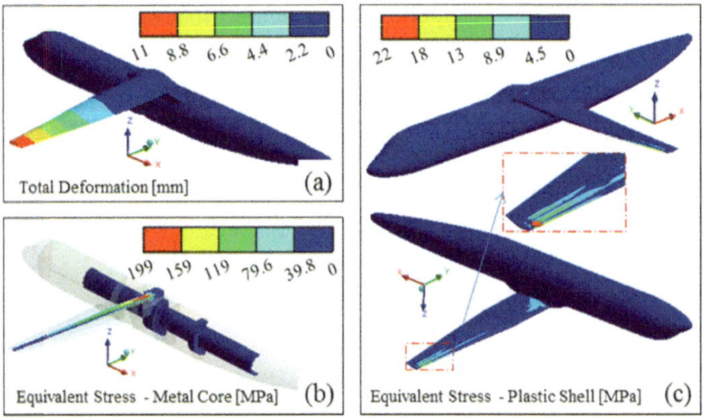

Fig. 8.9 Results of the strength calibration

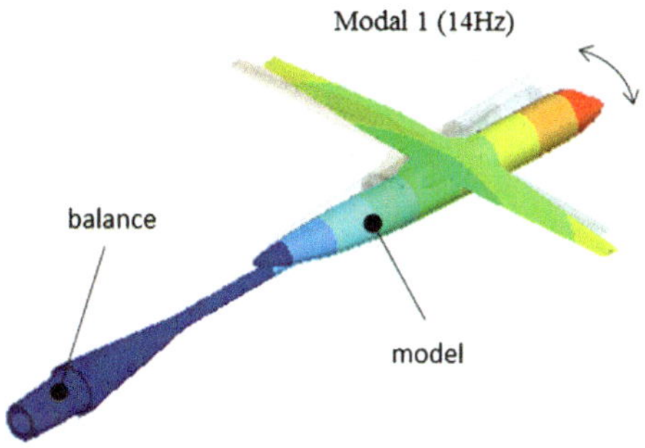

Fig. 8.10 First-order eigenmodes of the model-strut system

8.3.3 Resonance Checking

Figure 8.10 shows that the first-order modal frequency of the model-balance-strut system is 14 Hz, which avoids the blowing fundamental frequency of the FL-24 wind tunnel (10 Hz) and is safe for the test.

8.3.4 Stiffness Calibration Test

1) Test protocol

The calibration object is a plastic shell-metal beam assembly. This project evaluates the accuracy of the stiffness design by comparing the measured values of the model after loading deformation with the simulated calculated values. The deformation measurement is performed by a non-contact optical measurement system (XJTUSD, Xi'an Jiaotong University), which is based on digital close-up industrial photography and calculates the 3D coordinates of key points of the object by taking multiple images of the object before and after deformation to obtain data of the object in each state. And the multiple states are collocated with each other by global points or correlation points, and the deformation amount and deformation direction of the deformation points between each state are displayed by 3D chromatography.

Static deformation experimental platform as Fig. 8.11a, the measurement device is located above the test specimen (model wing). Figure 8.11b shows the clamping and loading scheme. The plastic housing is assembled to the metal frame and fixed to the table by a fixture located at the center axis of the fuselage; the fixture is made of aluminum, which is soft and can increase the contact area with the metal frame after being pretensioned to ensure a tight bond, so that the center axis of the model does not twist during loading. The weight block is set in the loading position of the wing through the metal frame; the metal frame is narrow, which can be approximated as a line load; there are four loading mark lines, as shown in Fig. 8.11c, and their coordinate positions are shown in Table 8.4.

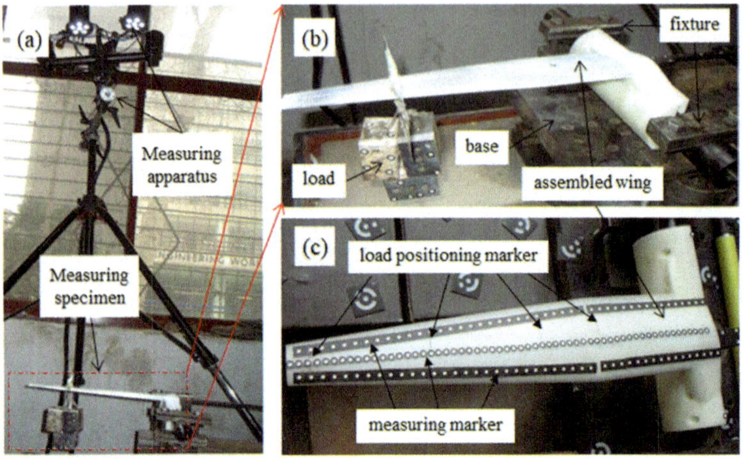

Fig. 8.11 Stiffness calibration test program

Table 8.4 Loading and deformation of the stiffness calibration test

Number	Loading		Maximum deformation	
	Location/mm	Load/N	Simulation value/mm	Test value/mm
1	152	57.48	1.5	1.7
2	200	57.48	2.8	3.1
3	300	26.65	3.6	3.3
4	400	26.65	7.9	8.0

2) Analysis of results

For the simulation calculations, the same modeling approach is used as for the strength check. The measured and calculated values of the maximum deformation at four loadings are shown in Table 8.4, and the values are in good agreement with each other and the test values are large.

The vertical component of the wing deformation during four loading times is shown in Fig. 8.12 along the spreading distribution, where the black point cloud is the experimental value, and the collection points include all three groups of marker points of leading edge, midline and trailing edge, and the red curve is the simulated value, and the spreading direction is collected along the rigid axis. In general, the test value is larger than the simulated value, and the stiffness of the finished model is smaller than the machined and assembled model. This may be caused by simulation modeling, design errors, material parameter deviations, differences in test boundary conditions and testing errors. However, the two are in the same trend of change and the maximum deviation is less than 5%, which is a good match and can meet the engineering needs.

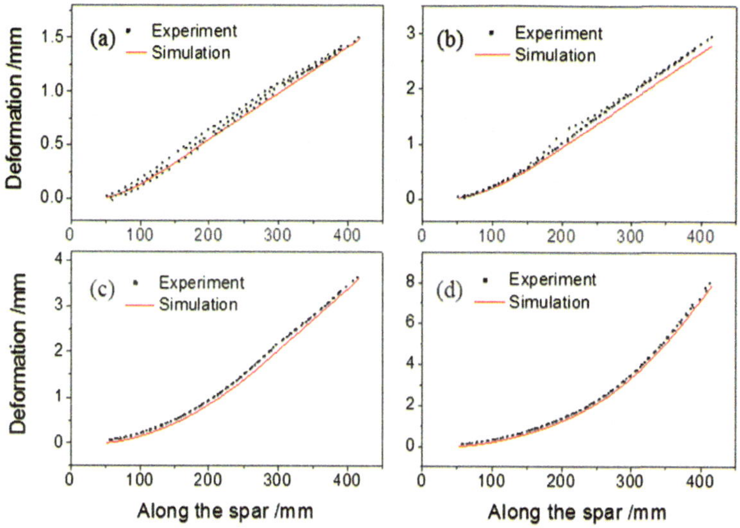

Fig. 8.12 Comparison of the results of the stiffness calibration test

In summary, it can be seen from the deformation test that the designed and processed model has high similarity of stiffness and meets the design requirements.

8.3.5 *Wind Tunnel Test*

The purpose of this test is to calibrate the static aeroelasticity calculations and to verify the theoretical calculation method. The data measured by the balance system is processed to derive its lift curve, and the static aeroelastic design is checked by observing the Mach number at the point of transition on the lift curve and comparing it with model-specific data.

The FL-24 transonic wind tunnel (test section size 1.2m × 1.2 m) was used; to simulate the real flight characteristics, the Mach number range was selected from 0.4 to 0.65 in steps of 0.05; the aerodynamic performance of the model was measured at a specific Mach number with an angle of attack from $-4°$ to $+6°$, with emphasis on the twisting phenomenon of the wing surface. The surface of the model was polished to reduce additional frictional drag; the model was assembled and the assembly holes on the surface were filled with putty; the model was attached to the connecting rod and fixed to the balance mechanism. The wind tunnel test program is shown in Fig. 8.13.

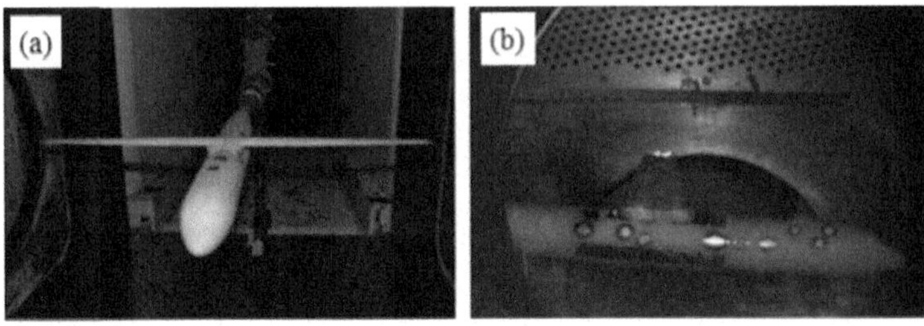

(a) Model installed in FL-24 wind tunnel (b) Test at an angle of attack of -2°

Fig. 8.13 Wind tunnel test program

8.4 Analysis and Discussion

8.4.1 Technology Availability

From the experimental phenomenon, it can be observed that the twist of the airfoil becomes more and more obvious with the increase of Mach number, and the subsequent data processing can be obtained that the lift-to-drag ratio curve has an obvious decreasing trend from Mach number 0.6–0.65, which is basically consistent with the twist dispersion rate of the actual vehicle.

For a flat wing with no operational rudder surface, the main phenomena of the aeroelastic static problem are torsional divergence and load redistribution, which can be analyzed using the classical binary airfoil segment model [90].

$$K_\theta - \frac{\partial C_L}{\partial \alpha} eqS = 0 \tag{8.9}$$

$$L = \frac{\partial C_L}{\partial \alpha}\left[\alpha_0 + \frac{\frac{\partial C_L}{\partial \alpha} eqS\alpha_0}{K_\theta - \frac{\partial C_L}{\partial \alpha} eqS}\right]qS = \frac{K_\theta \frac{\partial C_L}{\partial \alpha}}{K_\theta - \frac{\partial C_L}{\partial \alpha} eqS}\alpha_0 qS \tag{8.10}$$

Equation (8.9) is the torsional dispersion condition, and Eq. (8.10) is the lift expression when the wing undergoes elastic torsion. When the dispersion condition (8.9) is satisfied, the denominator of Eq. (8.10) tends to zero and the lift tends to infinity. And when the dynamic pressure q is smaller than the dispersion dynamic pressure, the denominator of (8.10) is greater than 0, and the lift force has a definite value, and the lift force will change with the dynamic pressure. A larger lift will result in a larger torsional moment and a more violent torsional deformation of the wing. When the unstable twist of the wing occurs, with the increase of the angle of approach, the airflow separation point of the surface layer of the wing moves forward and forms a larger turbulent area after separation, the aircraft lift is reduced and the differential pressure drag increases significantly. When the Mach number from 0.4 to 0.6, the lift coefficient C_L changes with the Mach number and the angle of approach, and the relationship between them is linear, when the Mach number changes from 0.6 to 0.65, although the lift coefficient also changes with the Mach number and the angle of approach, when the linear correlation decreases significantly. Especially, the drag coefficient C_D has a tendency to rise rapidly when the Mach number varies from 0.6 to 0.65, indicating that, at this time, the airfoil boundary layer separation point moves forward, forming a larger differential pressure drag. Also the relationship between the lift-to-drag ratio curve and Mach number can verify the existence of this process. In summary, it can be inferred that the torsional phenomenon intensifies between Mach number 0.6 and 0.65. The torsional dispersion velocity is roughly in this interval segment. The experimental simulation of a model with torsional dispersion velocity in this interval, and the experimental results are close to it. The feasibility of manufacturing a static aeroelastic model with similar stiffness based on light-curing printing technology was verified from the experimental point of view.

8.4.2 Technical Limitations

1) Subjects of the case study

The object of study is a regional airliner, 24.7 m long, 29.2 m wingspan, 8.85 m height, conventional aerodynamic layout, non-swept-back trapezoidal wing, T-shaped tail, design maximum takeoff weight 21,800 kg, maximum cruise speed 514 km/h, medium load, high requirements for flight safety; the model in this book is the airliner's wing model, excluding the T-shaped tail, rudder surface adjustment, and only simulates the wing stiffness; the test wind tunnel is the FL-24 transonic wind tunnel of China Aerodynamic Research and Development Center, with a test section size of 1.2 m × 1.2 m and medium size; the test Mach number range is from 0.4 to 0.65, and the angle of attack range is from $-4°$ to $+ 6°$, including the flight envelope. Therefore, the case study made in this chapter is typical and can be used as a reference for other tests of the same type.

2) Multi-material stiffness design method

The traditional metal beam-wooden frame segment (single or double beam) structure, the beam simulates the stiffness of the wing, and the stiffness of the wooden frame segment is negligible, therefore, the stiffness design of the model only needs to consider the structural parameters of the beam; the stiffness of the SL resin of the integrated aerodynamic shell is larger, and in order to achieve a higher accuracy of stiffness simulation, this book adopts the design method of resin-metal composite stiffness, and the stiffness calibration test shows that the composite structure designed by the The stiffness check test shows that the composite structure designed by this method can achieve high precision bending stiffness simulation. However, the resin-metal joint stiffness needs to be taken into account for the implementation of the complex elastic model using this method. In order to reduce the calculation difficulty of resin-metal connection stiffness, the structure design avoids glue joints, bolts and other firm but difficult to accurately simulate connections, which is feasible for the validation model with medium size and simple structure in this chapter, but in engineering practice, a large number of models have larger size, more detailed stiffness simulation and reuse many times, so there are higher requirements for the connection reliability between model parts that More complex interface processing is involved; from the perspective of optimization design, it actually increases the optimization objectives and optimization variables, and the design difficulty increases, requiring consideration of more complex model structures (such as increasing the cross-sectional parameters of metal beams) and more refined optimization algorithms.

Chapter 9
Additive Manufacturing of Wind Tunnel Test Models for Dynamic Aeroelastic Properties

9.1 Overview

Dynamically-scaled models (DSMs) are used in wind tunnel tests to study the flutter characteristics and other dynamic aeroelastic properties of aircraft or their components [67]. Dynamically-scaled models are a class of models used in wind tunnel tests. To facilitate the mapping of model data from wind tunnel tests to aircraft in real flight conditions, wind tunnels are designed and wind tunnel tests are conducted according to similarity laws [68]. The important similarity laws are as follows: (1) Geometry similarity: the similarity of model and prototype aerodynamic shape to ensure the similarity of flow field, applicable to all types of models; (2) Stiffness similarity: the similarity of model and prototype structural stiffness distribution. (2) Stiffness similarity: the similarity of stiffness distribution between the model and the prototype to ensure the similarity of the structural response of the model, which is applicable to the elastic model; (3) Mass similarity: the similarity of mass distribution between the model and the prototype to ensure the similarity of the mass characteristics of the model, which is applicable to the dynamic model. It is a common practice to select the most influential similarity criterion according to the test purpose, for example, the Static aeroelastic model does not need to satisfy the mass similarity; however, some types of models require high similarity and need to satisfy all the above similarity criteria, such as the dynamic similarity model [69]. However, there are some types of models that require high similarity and need to satisfy all the above similarity criteria, such as dynamic similarity models, in which the flutter test is used, called flutter models.

W. Zhu and D. Li, *Models for Wind Tunnel Tests Based on Additive Manufacturing Technology*, Additive Manufacturing Technology,
https://doi.org/10.1007/978-981-99-5877-1_9

The design and fabrication of dynamically similar models is usually based on metals, and the high modulus of the material makes it very difficult to ensure structural similarity while maintaining similar stiffness [70]. The high modulus of the material makes it very difficult to ensure structural similarity while maintaining similar stiffness. In order to prioritize the stiffness similarity of the model, the internal structure is usually simplified to neutralize the "overstiffness" caused by the high modulus of the model material, thus sacrificing the structural similarity. At the same time, the machining of metal models using traditional machining techniques (represented by CNC machines) requires long cycle times and high costs [6, 18]. The researcher has tried to use low-modulus metal models. Researchers have tried to use non-metallic materials with low modulus [70, 71], such as resin matrix composites, to design and fabricate dynamically similar models with partially similar structures. These studies have demonstrated that non-metallic materials can be used as an alternative to metallic materials to process dynamically similar, but without the support of advanced processing techniques, the design and processing complexity of the models are not effectively simplified, thus limiting the use of structurally similar modeling techniques. Therefore, it is necessary to follow the new advances in material science and manufacturing technology to find advanced technologies that can efficiently process complex low-modulus materials, so as to develop new methods for designing and manufacturing structurally similar dynamically similar models. However, these studies have focused on rigid wind tunnel test models, a class of models that only need to simulate aerodynamic shapes, while less research has been reported on the direct use of SL materials as a substitute for metals to machine dynamically similar models. In other fields outside of aerospace, researchers have used 3D printing to process all-plastic dynamically similar models to predict the dynamic properties of metal prototypes [72, 73]. Validation experiments have shown that plastic materials processed by the SL process can be approximately homogeneous, and their material properties are independent of the processing parameters (e.g., print orientation), which can be used to predict the dynamic properties of metallic structures [74]. The material properties are independent of the process parameters (e.g., print orientation) and can be used to predict the dynamic properties of metal structures.

In addition to satisfying similarity criteria such as geometric similarity, stiffness similarity and mass similarity, an ideal dynamically similar model should also simulate the force transmission path of the prototype, which requires structural similarity of internal force transmission components to ensure. Structural similarity is a broad concept that refers to the similarity of internal and external structures between the model and the prototype, where the external structural similarity is the similarity of the aerodynamic shape and the internal structure includes the similarity of structural stiffness, mass, and force transmission paths. Researchers have tried to use low modulus non-metallic materials [70, 71] to design and fabricate partially structurally similar dynamic similarity models. These studies have demonstrated that non-metallic materials can be used as an alternative to metallic materials to process the dynamic similarity, but the lack of support from advanced processing technologies has resulted in the failure to effectively simplify the design and processing complexity of the models, thus limiting the widespread use of structural similarity modeling techniques. On the one hand, the light-curing process SL has a strong ability to process complex internal and external structures, which is suitable for the fabrication of wind tunnel products with complex shapes and internal structures (complex force transfer structures); on the other hand, the plastic materials processed have small elastic modulus and sufficient strength, which provides the possibility of structural similarity and stiffness similarity in the design of dynamically similar models.

This chapter proposes a new method for the design and fabrication of structurally similar dynamically similar models based on additive manufacturing technology, which is capable of processing models with similarity in shape, stiffness, and internal structure, while effectively reducing the cycle time and cost of wind tunnel test model design and fabrication with the high efficiency of light-curing printing technology processing. The chapter will detail the principles and key technologies of the new method and use a case study to demonstrate the feasibility of this technology.

The technical scheme of this chapter is shown in Fig. 9.1. The dynamic characteristics of the aluminum alloy prototype were obtained by the finite unit method calculation, and the target data were obtained after the scaling design. In order to verify the validity of the model design and processing, a modal test was conducted on the resin model, and the actual data obtained from the test was compared with the target data to determine the degree of agreement between the two. At the same time, structural analysis was performed to verify the structural similarity between the prototype and the model; finally, the economics of the new method of model design and fabrication were also discussed.

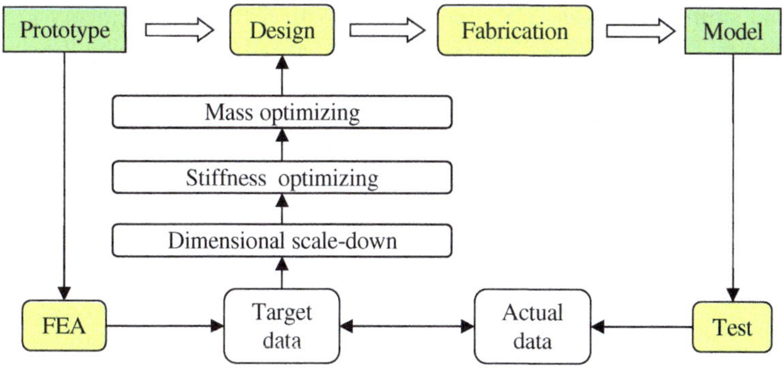

Fig. 9.1 Technical solution and model design flow

9.2 Design of Structural Similarity Model

As mentioned above, this section completes the design of a dynamically similar model with similar structure through three steps: dimensional scaling, stiffness optimization and mass optimization, with metal as the prototype material and plastic as the model material.

9.2.1 Size Scaling

This is the first step of the model design to obtain the initial structure and the exact aerodynamic shape of the internal structure of the model. The scaled external and internal dimensions of the prototype are scaled according to the geometric similarity coefficient, and the scaled external dimensions define the aerodynamic shape of the model and remain unchanged in the subsequent design to ensure "geometric similarity"; the scaled internal dimensions define the initial state of the internal structure of the model, and the final dimensions are determined in the stiffness optimization and mass optimization. The scaled internal dimensions define the initial state of the internal structure of the model, and the final dimensions will be determined in the stiffness optimization and mass optimization, but the subsequent design keeps the topology of the initial state unchanged and only changes the size to ensure "structural similarity".

Table 9.1 lists the similar factors that need to be satisfied in the design of the dynamic similarity model. They are derived from the similarity theory [75]. The latter can be calculated based on the former, and the key to determine the similarity criteria is the basic criteria [19]. The key to determine the similarity criterion is the basic criterion. The dimensional similarity criterion is the scaling factor of the prototype and model structure size, which is the key parameter of the scaling design and is determined according to the prototype size, wind tunnel test section size and other parameters; the medium velocity and density similarity criterion is determined according to the real flight state and wind tunnel fluid related parameters; the Mach number Ma similarity criterion is 1:1 according to the similarity theory.

9.2.2 Stiffness Optimization

This book adopts the Matrices of Elastic Influence Coefficients (MEIC) method for the design of model stiffness [69]. The stiffness is the deformation capacity of a structure to withstand a load. Stiffness is the deformation capacity of a structure to withstand a load, and its magnitude is measured by the deformation of the structure

Table 9.1 Similarity criterion for dynamic similarity model

Similarity criteria		Expressions
Basic guidelines	Size	K_l
	Fluid velocity	K_v
	Fluid density	K_ρ
	Mach number	K_{Ma}
Guidelines for derivation	Stiffness	$K_{EI}/K_{GJ} = K_\rho K_v^2 K_l^4$
	Quality	$K_M = K_\rho K_l^3$
	Load	$K_F = K_\rho K_v^2 K_l^2$
	Deformation	$K_\delta = K_l$
	Frequency	$K_\omega = K_v / K_l$

under a certain load. Therefore, Stiffness similarity can be converted into Displacement similarity for a specific direction under similar loads. The specific direction is determined according to the dynamic structural properties of the model, depending on the modalities of the first few orders of the model.

The stiffness design is done automatically using optimization calculations with the following mathematical model:

$$\begin{cases} \text{minimize} & f(\boldsymbol{x}) \\ \text{subject to} & \boldsymbol{x}^l \leq x \leq \boldsymbol{x}^u \end{cases} \tag{9.1}$$

1) Objective function

As mentioned above, similar stiffness is defined as similar deformation under similar loads, and therefore the deformation error is used as the objective function for optimization:

$$f(x) = \sum_{i=1}^{m} \sum_{j=1}^{n} \left[\frac{\delta_{ij}(x) - \overline{\delta}_{ij}}{\overline{\delta}_{ij}} \right]^2 \tag{9.2}$$

where m is the number of loadings, and n is the number of deformation calculation nodes, the δ and $\overline{\delta}$ are the calculated and target deformations at a node under a particular load, respectively, and the latter can be calculated by the following equation:

$$\overline{\delta} = \delta_P \cdot K_l \tag{9.3}$$

where δ_P is the amount of deformation of the prototype (Prototype) at a node, and K_l is the deformation similarity criterion.

2) Optimization variables

The optimization variables are the internal structural dimensions of the model, which define the size of the internal structure, while its topology (e.g., number of webs, ribs, boom trusses and interconnections) remains unchanged; the external dimensions (length, width, height) of the model do not change during the optimization process (already determined in the dimensional scaling stage) to maintain geometric similarity.

3) Control variables

The control variables are set to make the optimization results reasonable and processable.

9.2.3 Quality Optimization

The motion control equation is as Eq:

$$[M]\{\ddot{q}(t)\} + [C]\{\dot{q}(t)\} + [K]\{q(t)\} = F(t) \tag{9.4}$$

where $[M]$, $[C]$ and $[K]$ are the generalized mass matrix, the generalized damping matrix and the generalized stiffness matrix; *F(t)* is the generalized load vector, which is zero for modal solutions; $\{\ddot{q}(t)\}$, $\{\dot{q}(t)\}$ and $\{q(t)\}$ are the generalized acceleration vector, generalized velocity and generalized coordinates, respectively.

The mass properties of the model refer to the total weight, center of mass position and rotational inertia of the model. One way of solving is to solve the model mass characteristics directly according to the prototype-related parameters and similarity criterion (as shown in Table 9.1); another way is to solve it indirectly according to the model structural characteristics. As shown in Eq. (9.2), the mass properties, damping properties and stiffness properties jointly determine the dynamic properties of the model According to the above discussion, the effect of damping properties can be neglected in this book, and Eq. (9.2) is simplified as follows

$$[M]\{\ddot{q}(t)\} + [K]\{q(t)\} = F(t) \tag{9.5}$$

That is, the mass characteristics and stiffness characteristics jointly determine the model dynamic characteristics, and the model dynamic characteristics can be characterized by the Eigen mode of the model. The eigenmodes of the model can be obtained from the eigenmodes of the prototype wing box segment with the corresponding similarity criterion, while the stiffness characteristics of the model have been obtained from the stiffness optimization, so the mass characteristics can be inverse obtained from the known dynamic characteristics and stiffness characteristics. Based on this, the mathematical model for mass optimization is established as follows:

$$\begin{cases} \text{minimize} & g(\boldsymbol{x}) \\ \text{subject to} & \boldsymbol{x}^l \leq x \leq \boldsymbol{x}^u \end{cases} \tag{9.6}$$

1) Objective function

As mentioned above, the model mass characteristics are derived from the model dynamic characteristics and stiffness characteristics. In the mass optimization based on the above stiffness optimization, the resulting structure is used as the input data for this optimization and remains unchanged, and the stiffness characteristics are already reflected; as for the dynamic characteristics of the model, they are obtained from the characteristic modes (including characteristic frequencies and characteristic vibration patterns) of the prototype wing box section and the corresponding similarity criterion, and the vibration pattern similarity is naturally satisfied due to the similarity between the model and the prototype structure, and only the characteristic frequencies need to be considered. Therefore, the objective function can be defined as the error between the actual eigenfrequency of the model and the target frequency as follows:

$$g(\mathrm{x}) = \sum_{j=1}^{\mathrm{m}} \left[\frac{\omega_{\mathrm{j}}(\mathrm{x}) - \overline{\omega}_{\mathrm{j}}}{\overline{\omega}_{\mathrm{j}}} \right]^2 \tag{9.7}$$

where m is the order of the considered eigenfrequencies, the ω and $\overline{\omega}$ are the actual and target frequencies of the model, respectively, the latter being obtained from the eigenfrequencies of the prototype (Prototype) and the frequency similarity criterion:

$$\overline{\omega} = \omega_P \cdot \mathrm{K}_{\omega} \tag{9.8}$$

where ω_P is the eigenfrequency of a certain order of the prototype, and K_{ω} is the eigenfrequency similarity criterion.

2) Optimization variables

This book adjusts the mass characteristics of the model by means of the counterweight blocks (balance-weight). Where, the total weight of all the counterweight blocks M_{b} is obtained from the following equation:

$$M_{\mathrm{b}} = \overline{M} - M_0 \tag{9.9}$$

where M_0 is the initial total weight of the model, which is the total amount of the model after the above stiffness optimization; $\overline{M}$ is the target total weight of the model, which is determined by the total weight of the prototype M_P and the mass similarity criterion K_M are obtained as follows:

$$\overline{M} = M_P \cdot \mathrm{K}_M \tag{9.10}$$

The number, position and weight of individual counterweight blocks can be used as optimization variables.

3) Control variables

The control variables are set to make the optimization results reasonable and processable.

9.3 Case Studies

9.3.1 Wing Prototype

In this book, the wing box segment is selected for the study.

In order to approximate the real wing as much as possible, the book is designed as Fig. 9.2. The prototype is a whole structure made of aluminum alloy, including the main load-bearing parts of the wing: skin, ribs, boom and web, etc. The materials used in the prototype and the model are aluminum alloy and SL plastic respectively, and their material properties are shown in Table 9.3.

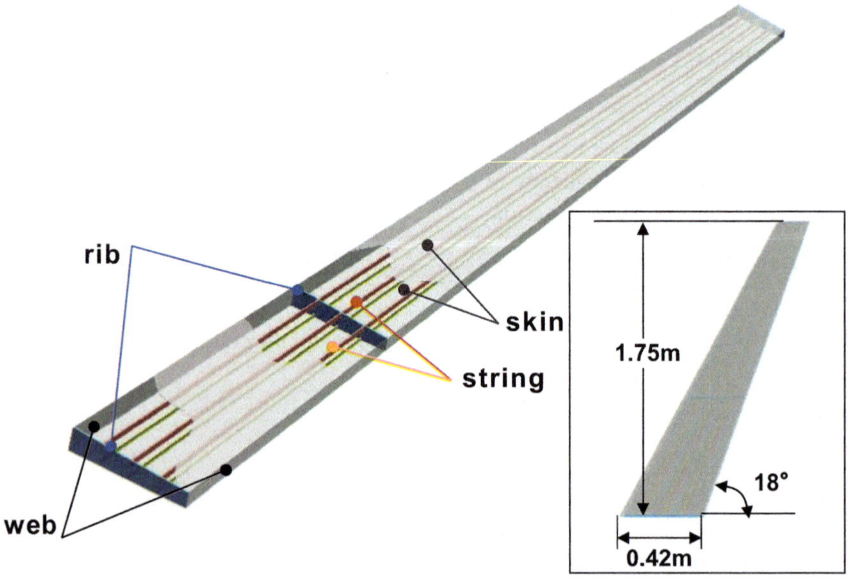

Fig. 9.2 Schematic diagram of the structure of a typical wing box section

9.3.2 *Optimized Design*

1) Size Scaling

The proposed test condition in this book is subsonic speed. The size similarity criterion is the scaling factor of the prototype and model structure size, which is the key parameter of the scaling design, and is determined according to the prototype size, wind tunnel test section size and other parameters, and 1:5 is selected in this book; the similarity criterion of medium velocity and density is determined according to the real flight state and wind tunnel fluid related parameters, and 1:4 and 1:1 in this book respectively; according to the similarity theory, the Mach number *Ma* similarity criterion is 1:1, as shown in Table 9.2.

2) Stiffness optimization

Based on the above discussion, the mathematical model for stiffness optimization developed in this book is as follows:

Table 9.2 Similarity criteria for dynamic similarity models

Similarity Criteria		Expressions	Value[a]
Basic Guidelines	Size	K_l	1:5
	Fluid velocity	K_v	1:4
	Fluid Density	K_ρ	1:1
	Mach number	K_{Ma}	1:1
Guidelines for derivation	Stiffness	$K_{EI}/K_{GJ} = K_\rho K_v^2 K_l^4$	1:10,000
	Quality	$K_M = K_\rho K_l^3$	1:125
	Load	$K_F = K_\rho K_v^2 K_l^2$	1:400
	Deformation	$K_\delta = K_l$	1:5
	Frequency	$K_\omega = K_v / K_l$	1:1.25

[a]The similarity ratio is model-prototype

Table 9.3 Results of model quality optimization

Parameters		Error/%
Total weight	M	−1.0
Position of the center of mass	x	−6.6
	y	−3.1
	z	13.6
Rotational inertia	Ix	4.4
	Iy	−2.3
	Iz	−2.3

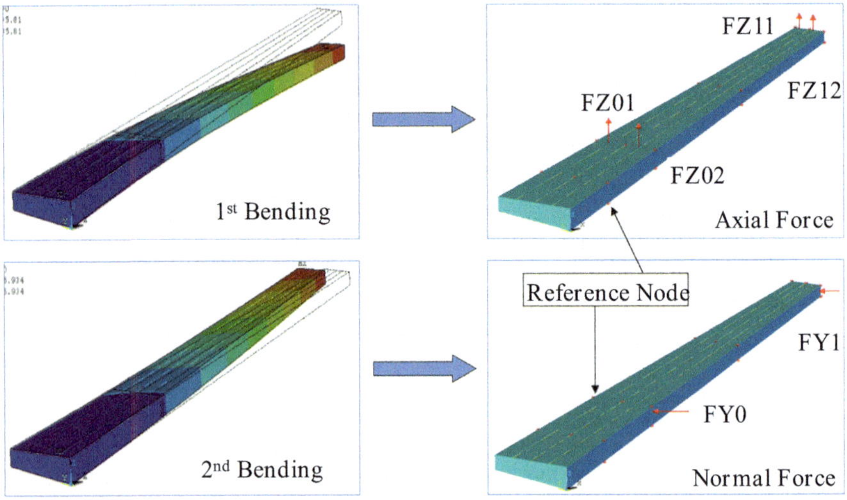

Fig. 9.3 Principle of stiffness design using the elastic influence coefficient matrix method

(1) Objective function

As Fig. 9.3 shown, the first and second order modes of the prototype are upper and lower bending and chordal bending, respectively, and the load can be set and the deformation compared along these two directions.

Using the deformation error as the objective function for optimization:

$$f(x) = \sum_{i=1}^{m} \sum_{j=1}^{n} \left[\frac{\delta_{ij}(x) - \overline{\delta}_{ij}}{\overline{\delta}_{ij}} \right]^2 \tag{9.11}$$

where m is the number of loads (2 in Fig. 9.2), and n is the number of nodes for deformation calculation (24 in total as shown by * in Fig. 9.2), the δ and $\overline{\delta}$ are the calculated and target deformations at a node under a specific load, respectively, and the latter can be calculated by the following equation:

$$\overline{\delta} = \delta_P \cdot K_l \tag{9.12}$$

where δ_P is the amount of deformation of the prototype (Prototype) at a node, and K_δ is the deformation similarity criterion.

(2) Optimization variables

As Fig. 9.4 shown, the optimization variables are the internal structural dimensions of the model, such as rib thickness (root, middle and slightly, 3 in total), skin thickness (upper surface, lower surface, 2 in total), web thickness (leading edge, trailing edge,

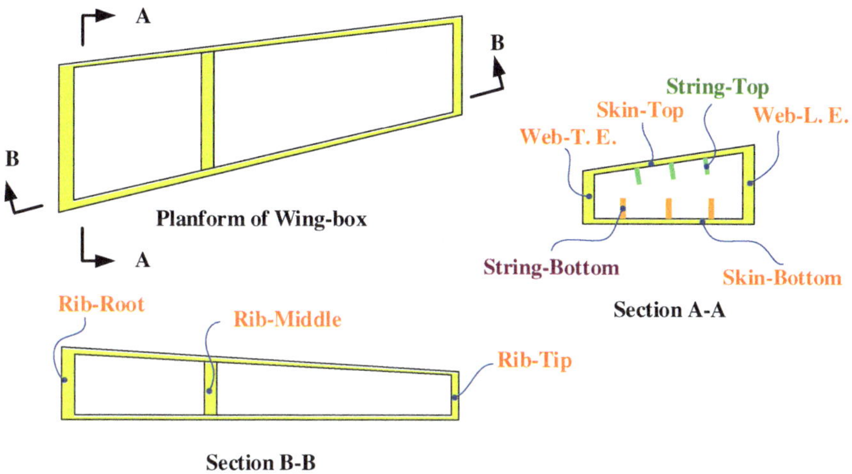

Fig. 9.4 Distribution of optimization variables

2 in total) and the height and width of the boom truss (root upper surface, root lower surface, middle upper surface, middle lower surface, slightly upper surface, slightly lower surface, 12 in one).

(3) Control variables

Considering that the dimensions of the root skin, ribs, boom truss, web, etc. are larger than the slightest part, the minimum value of all dimensions should be larger than the minimum value allowed by the SL printing process (0.5 mm), etc.

In this book, BEAMP181 element in ANSYS® is used to build a finite element model of the stiffness of the wing box section, and its optimization module is used to build an automatic design to implement Single Run, DV Sweep, and First-order optimization algorithms to find the optimum.

3) Quality optimization

The quality optimization model for this book is based on the above approach as follows:

(1) Objective function

The objective function is defined as the error between the actual characteristic frequency of the model and the target frequency:

$$g(\mathrm{x}) = \sum_{j=1}^{\mathrm{m}} \left[\frac{\omega_{\mathrm{j}}(\mathrm{x}) - \overline{\omega}_{\mathrm{j}}}{\overline{\omega}_{\mathrm{j}}} \right]^2 \tag{9.13}$$

where m is the order of the eigenfrequencies considered (e.g., only 1st order bending and 2nd order bending are considered in this book, so $m = 2$), the ω and $\overline{\omega}$ are the actual and target frequencies of the model, respectively, the latter being obtained from the eigenfrequencies of the Prototype and the frequency similarity criterion:

$$\overline{\omega} = \omega_P \cdot \mathrm{K}_{\omega} \tag{9.14}$$

where ω_P is the eigenfrequency of a certain order of the prototype, and K_ω is the eigenfrequency similarity criterion.

(2) Optimization variables

This book adjusts the mass characteristics of the model by means of the counterweight blocks (balance-weight). Where, the total weight of all the counterweight blocks M_b is obtained from the following equation:

$$M_\mathrm{b} = \overline{M} - M_0 \tag{9.15}$$

where M_0 is the initial total weight of the model, which is the total amount of the model after the above stiffness optimization; $\overline{M}$ is the target total weight of the model, which is determined by the total weight of the prototype M_P and the mass similarity criterion K_M are obtained as follows:

$$\overline{M} = M_P \cdot \mathrm{K}_M \tag{9.16}$$

Considering the different load-bearing capacity of each structural part of the model (the counterweights should be installed at locations with high load-bearing capacity, such as the wing roots), the nodal line location of the target characteristic vibration pattern (the counterweights should be installed to avoid the extreme value of the vibration pattern nodal line to reduce the influence on the vibration pattern) and the convenience of the assembly operation (the counterweights should be installed at convenient locations for disassembly and assembly, such as the surface of the wing box), the book has selected the following Fig. 9.3 The 12 locations shown in Fig. 9.3 are chosen as the mounting points of the counterweight. Therefore, the number of counterweights and their locations have been determined, using the ratio of the weight of individual counterweights to the total counterweight weight as the optimization variable:

$$\boldsymbol{x} = (x_1, x_2, \ldots, x_{12})^{\boldsymbol{T}} \tag{9.17}$$

where $x_1, x_2, \ldots, x_{12}$ is the ratio of each counterweight block in the total counterweight, and takes a value in the range of (0, 1).

(3) Control variables

This book sets the minimum weight of a single counterweight block to 0.5g, and the maximum value is determined by the design and is not set separately.

(4) Optimize implementation

Similar to the stiffness optimization, the MASS21 element in ANSYS® is used to build the mass finite element model of the wing box section, and its optimization module is used to build the automatic design, which implements the Single Run, DV Sweep and First-order optimization algorithms to find the optimum. The quality characteristics of the optimized parameters are shown in Table 9.3.

9.3.3 Model Processing

1) Model body structure

The main structure of the model refers to the load-bearing structure of the model, which is the result of the stiffness optimization design and does not include the counterweight section. The designed wing box section was processed by SL process with SPS600B equipment and SOMOS® 14,120 (DSM Corp.) material. The CAD data of the model was created according to the optimized design results, the SL process parameters were selected, and the model was imported into the SL machine for processing, and the required solid model was obtained after cleaning and grinding.

The fine dimensions of the main structure of the model are shown in Fig. 9.5 They are mainly concentrated in the wing tip area, including the thickness of the trailing edge web (web-T. E.), the thickness of the upper skin (skin-top), the thickness of the wing tip rib (rib-tip), the width and height of the bottom boom truss (string-bottom), the thickness of the leading edge (web-L. E.) and the thickness of the lower skin (skin-bottom), etc. The smallest dimension is 0.5 mm for the boom truss width, which is within the printing capability of the SL process (the minimum processing size is about 0.5 mm). As shown in the figure, most of these structures are located in the closed inner cavity of the wing box, which is difficult to achieve by mechanical processing methods and requires the SL technology with internal structure processing capability.

The selection of process parameters is a key technology for SL processing, in which most of the process parameters are selected with preset values (such as layered thickness, laser power, scanning speed, etc.), and the selection of printing direction needs to be focused on. Because the model in this book is a cavity structure, too much internal process support will bring about the loss of dimensional accuracy and surface quality of the inner cavity, such as the fine structure of the wing tip area (as Fig. 9.5), therefore, the internal support needs to be reduced or eliminated by optimizing the printing direction.

As shown in Fig. 9.6 the internal support is a process structure used to prevent the top structure of the part from deforming or even collapsing due to gravity. For cavity structures that require high internal surface accuracy and are not easily post-processed, the internal support needs to be reduced or even avoided altogether by optimizing the print orientation. For slender cavity structures with an advantageous length, increasing the print angle (the angle between the length of the part and the

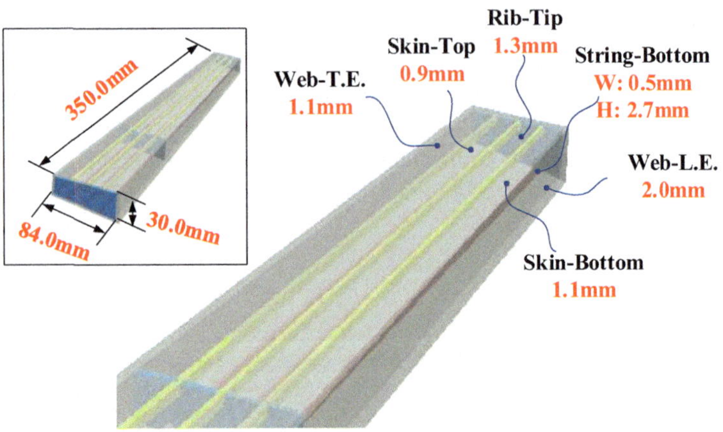

Fig. 9.5 Extreme size distribution of the model

surface of the machining platform) reduces the internal support required for the top structure, but increases the printing time due to the elongation of the print height. Therefore, the optimal print orientation needs to be chosen to minimize the internal support while keeping the print time to a minimum.

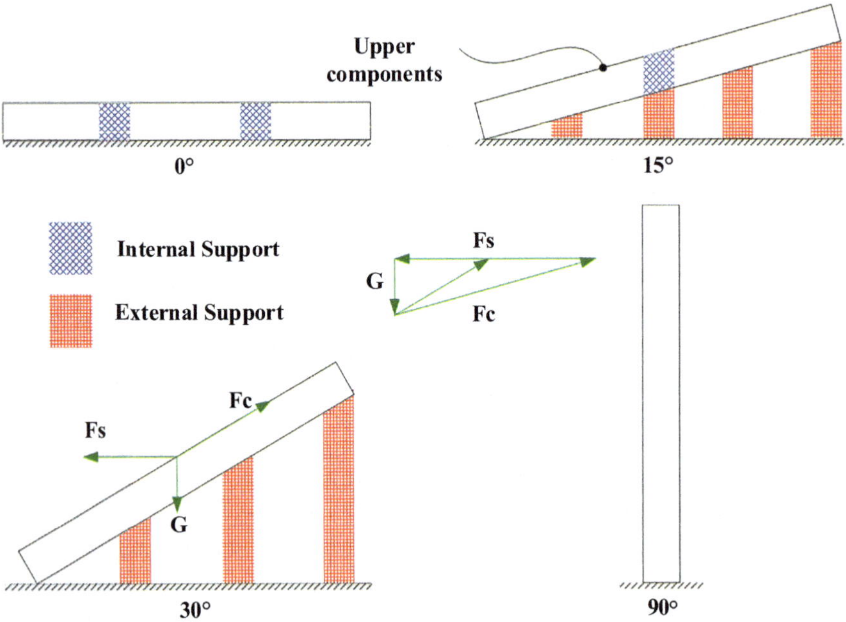

Fig. 9.6 Relationship between SL printing direction and internal and external support

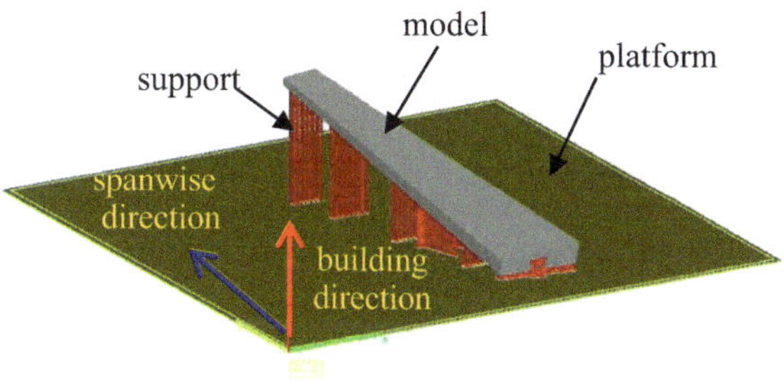

Fig. 9.7 Printing direction of SL process machining model

This book uses the above method to select the optimal printing direction to ensure that there is no internal process support, and the printing direction is about 30° from the wingspan angle, as Fig. 9.7 shown.

2) Counterweight structure

As shown in Fig. 9.8, the counterweight material is lead, and the positioning structure is printed directly by SL and installed by 502 glue joint. The choice of counterweight material is examined in terms of density, processability, elastic modulus and price. In order to reduce the impact on the stiffness of the box section members, the density of the counterweight material should be large enough to reduce the assembly surface, while its elastic modulus should be as close as possible to that of the main structure; at the same time, the counterweight material should also have good processability and low cost. Lead has a large density (11.34 g/cm^3, second only to tungsten among the commonly used high-density materials), and its modulus of elasticity is close to that of SL plastic (16 GPa) compared with other high-density materials (tungsten, brass, etc.), and it has better processability and lower cost, therefore, lead is chosen as the counterweight material in this book.

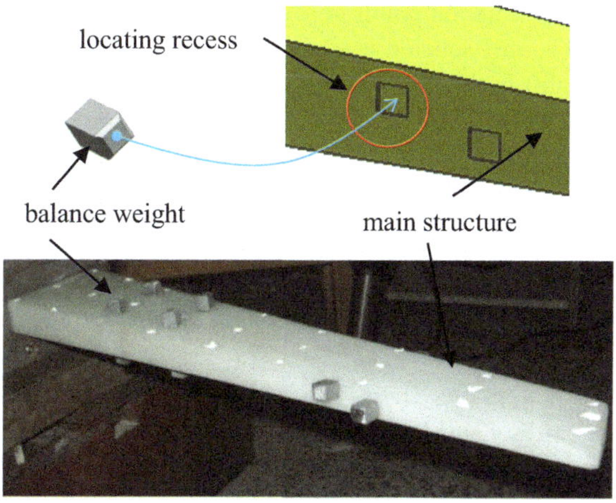

Fig. 9.8 Weighting scheme of the model

The counterweight block is the carrier for providing counterweight. The design of counterweight block structure not only meets the counterweight requirements, but also takes into account the processability and ensures the connection with the box section model. For the convenience of processing, the counterweight block is designed as a hexahedral block structure, and the connection is made by using the counterweight block inlay positioning glue bonding scheme, and the connection surface is set to insert the tab structure, as shown in Fig. 9.8. The positioning accuracy of the counterweight has an important impact on the simulation accuracy and is a difficult point for the traditional fluttering model manufacturing, but the SL technology provides a new solution for the high-precision positioning structure. In this book, the ANSYS optimized coordinates of the counterweight are imported into CAD design software, and the positioning structure is designed with this in mind (as shown in Fig. 9.8), and finally, the counterweight positioning structure is manufactured with high precision by direct manufacturing with SL. Due to the pre-fabrication of the assembly surface in the main structure, the installation of the counterweight is also greatly simplified.

9.3.4 Verification of Stiffness Characteristics

The accuracy of the stiffness design is evaluated by comparing the measured values of the model after loading deformation with the simulated calculated values. The deformation measurement is performed by a non-contact optical measurement measurement system (XJTUSD, Xi'an, China), which is based on digital close-up industrial photography and calculates the 3D coordinates of key points of the object by taking multiple images of the object before and after deformation to obtain data of the object in each state. And the multiple states are combined with each other by global points or related points, and the deformation amount and deformation direction of the deformation points between each state are displayed by 3D chromatography.

The test scheme is shown in Fig. 9.9. The machined model (without counterweight) is fixed on the measuring table by a custom-made fixture, and a weight with a load of 20 N is fixed on the slight part of the model by an iron hook, and the measuring points are distributed as follows Fig. 9.9. Figure 9.9b shows the simulation analysis model.

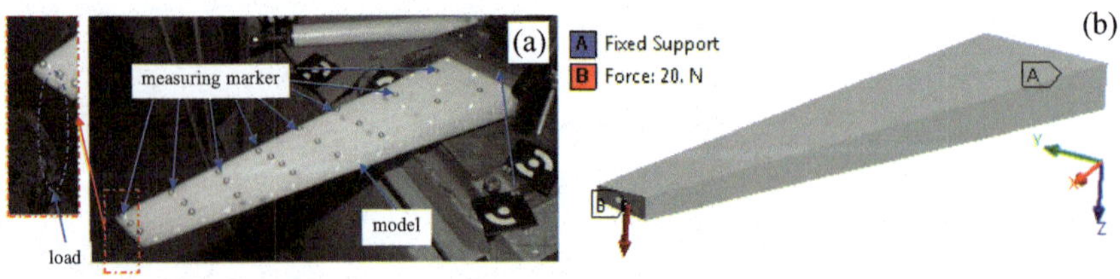

Fig. 9.9 Validation scheme

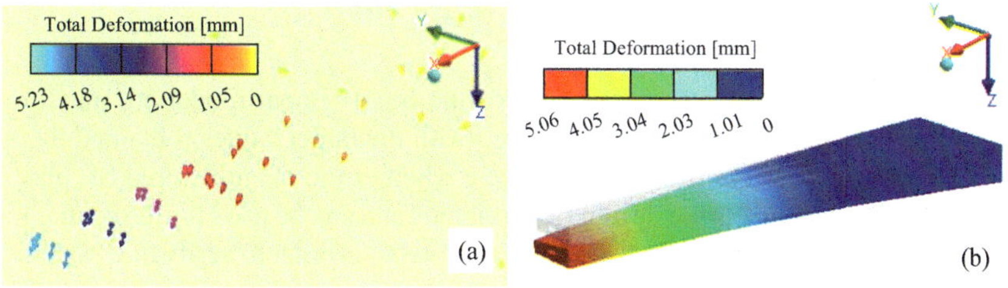

Fig. 9.10 Validation results

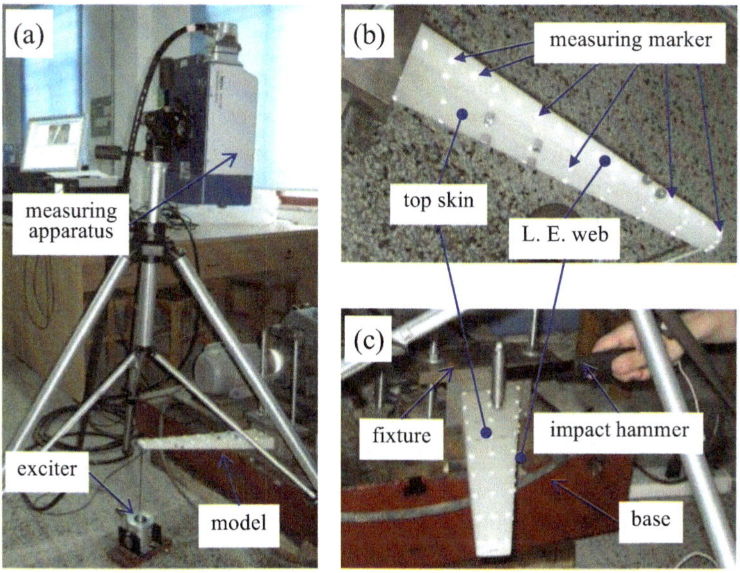

Fig. 9.11 Modal experimental scheme of the wing box segment model

The measured and simulated values of model deformation under the same working conditions are shown in Fig. 9.10a, b show. It can be easily obtained that the deformation distribution is consistent, and the maximum deformation occurs at the slight part, and the error is small (about 3%), so it can be inferred that the stiffness distribution of the finished model matches with the expectation and the SL has high machining accuracy.

9.3.5 Dynamic Characterization Verification

In this book, the dynamic characteristics of the box segment model are measured by modal tests, and the results are compared with the target values (obtained by the prototype and similarity criterion) to analyze whether the manufactured SL plastic box segment model meets the design requirements and can be used for the estimation of the characteristic parameters of the metal prototype, in order to verify the feasibility of the method proposed in this book. Based on a scanning laser vibrometer (PSV-400, Polytec) with non-contact measurement, the book establishes the following. The plastic box segment is mounted on the measurement base with a customized fixture, and the measuring marker is pre-set on the top skin and L. E. web of the box segment. The measuring apparatus is set up on the upper part of the model to scan the measuring points and obtain the dynamic characteristics (characteristic frequencies and characteristic vibration patterns in this book) of the SL plastic box section after processing (FFT, etc.).

The results of the modal measurements are sensitive to the boundary conditions, including the assembly of the specimen components and the fixed mounting of the specimen. For the fixed mounting of the specimen, a steel fixture was designed as shown in Fig. 9.12. The fixture is designed to ensure the fixed connection of the plastic cassette at the wing root on the one hand, and to allow the top skin and L. E. web of the cassette to face upward vertically towards the scanner head of the vibrometer after mounting to the measurement substrate on the other hand.

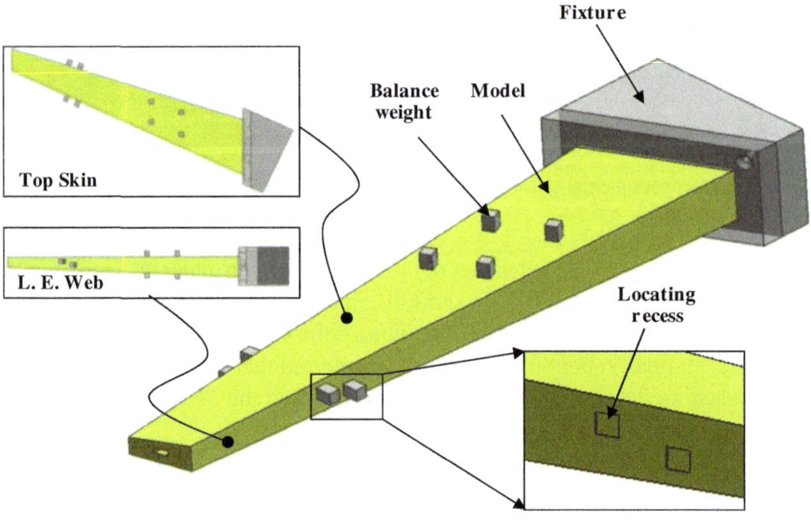

Fig. 9.12 Part assembly and fixture mounting of the model

As mentioned above, the plastic wing box section of this specimen includes two types of components: the main structure and the counterweight block. Since the load-bearing capacity of each part of the main structure was considered in the design of the counterweight position and the extreme value point of the first 2 orders of characteristic vibration pattern was avoided, the design requirements were satisfied by using 502 instantaneous dry glue to assemble and connect the parts. The corresponding counterweights are fixed in the way shown in Figs. 9.8 and 9.12.

The measured time domain vibration signal is transformed to frequency by FFT to obtain the dynamic characteristics of the box section. Figure 9.13 gives the first 2 orders of eigenfrequencies and eigenvibrations of the plastic box section. The comparison results of the eigenfrequencies are shown in Table 9.4, where the actual values are obtained from this experiment, and the target values are obtained from the characteristic parameters of the aluminum box segment and the similar criterion.

The first 2 orders of characteristic vibration patterns of the box section are up and down bending oscillation and left and right bending oscillation, which are obtained by scanning the upper skin and the leading edge web, respectively, from Fig. 9.13 The actual and target values of the vibration patterns are similar, as can be seen in Table 9.4 It is easy to obtain that the actual and target values are in good agreement. The measurement of the SL plastic box section can predict the dynamic characteristics

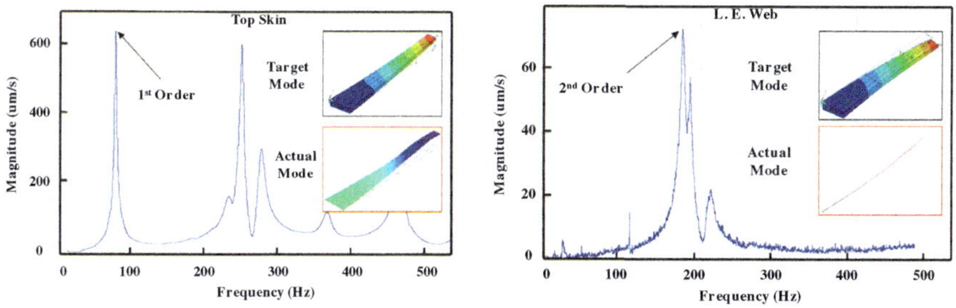

(a) Measured from the top skin (b) Measured from the leading edge web

Fig. 9.13 Experimental results of plastic box segment model states

Table 9.4 Comparison of thc target and actual values of the characteristic frequency of the plastic box section

Order	Analysis/Hz	Experiment/Hz	Error/%
1	69.58	70.63	−1.49
2	177.00	176.90	0.06
3	255.37	274.40	−6.94
4	352.91	366.00	−3.58
5	451.72	465.00	−2.86

of the metal prototype box section more accurately, which lays the foundation for obtaining the dynamic aeroelastic parameters of the metal prototype aircraft in the wind tunnel [44, 46].

9.4 Analysis and Discussion

9.4.1 *Structural Similarity Analysis*

Figure 9.14 shows the internal structure of the aluminum box segment prototype and the plastic box segment model.

By comparing the internal structures of the aluminum box section prototype and the plastic box section model, it can be seen that both have the same topology and similar force transfer characteristics, so the power similar resin model designed in this book achieves structural similarity. In contrast, the metal dynamically similar model requires a lot of simplification of the internal structure of the prototype, which makes it difficult to achieve structural similarity [19]. Therefore, the proposed design and fabrication method of plastic dynamically similar model based on SL process in this book can make the simulation degree of dynamically similar model further compared with the traditional metal mechanical processing method.

Compared to metals, resins have a smaller modulus of elasticity, and the implications of this for model design are discussed below. Without loss of generality, a hollow rectangular beam is chosen as the object in this book to analyze the reasons

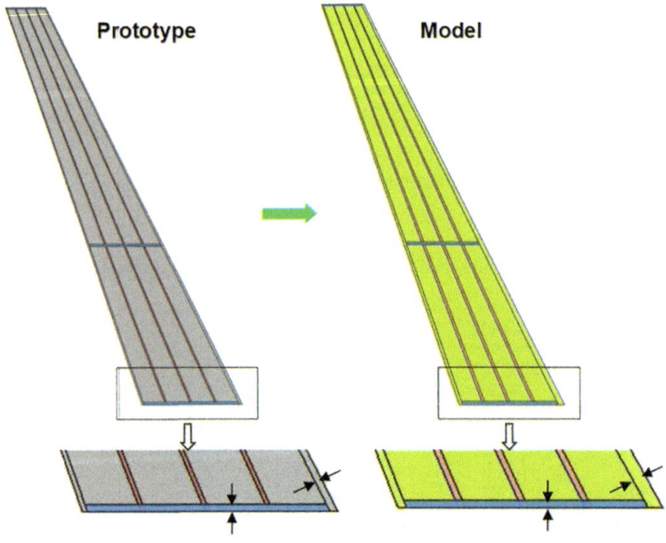

Fig. 9.14 Internal structure of the aluminum box segment prototype and plastic box segment model

why the resin model can achieve structural similarity. The structure and dimensions of the prototype beam are shown in and include five parameters: length (l), width (w), height (h), and wall thickness in the width direction (thick-X, tx) and height direction (thick-Y, ty).

According to the design method of dynamic similar model, the hollow rectangular beam dynamic similar models of SL plastic and aluminum alloy are designed with w, tx and ty as optimization variables (h and l scaling ratio are designed to be unchanged to keep geometric similarity), and their structures and dimensions are shown in and respectively.

1) Similar structure

From Fig. 9.15 and Table 9.5 it can be seen that the SL resin model has the same topology as the prototype, and at the same time the structural dimensions are machinable, so the structural similarity of the dynamically similar model can be achieved based on the SL process. On the contrary, the minimum wall thickness ty = 0.05 mm of the aluminum alloy model obtained by the structural similarity design is not economically machinable, and in order to realize the dynamic similarity, only the internal structure, such as the I-beam, can be changed, thus losing the structural similarity. Taking the bending stiffness EI As an example, the reason for this result is argued. According to the similarity theory, the stiffness similarity criterion K_{EI} can be expressed as follows

$$K_{EI} = K_E \cdot K_I \tag{9.18}$$

From the above equation, if the stiffness similarity coefficient is certain, the similarity coefficient of the material stiffness parameter (here is the modulus of elasticity

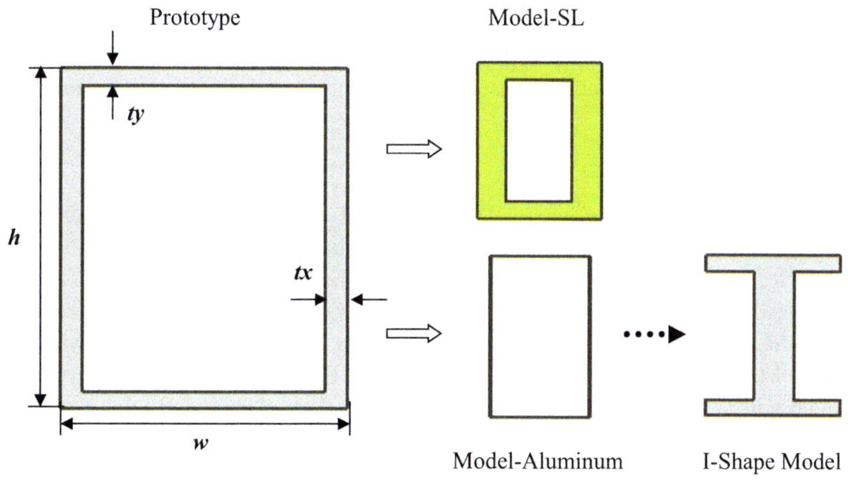

Fig. 9.15 Cross section of the prototype and model rectangular hollow beam

Table 9.5 Parameters of the prototype and model rectangular hollow beam

Parameters		Prototype	Models—Resin	Model—Aluminum
Size/mm	Length	400.00	80.00	80.00
	Height	100.00	20.00	20.00
	Width	80.00	15.14	11.78
	Thickness-X	6.00	3.53	0.12
	Thickness-Y	5.00	2.14	0.05
Mass/g	Main structure	2090.6	15.6	1.3
	Counterweight	16.7[a]	1.1 (0.1)[b]	15.4 (11.9)[b]

[a] The target mass of the model is derived from the prototype mass and the mass similarity criterion
[b] Mass ratio of counterweight to main structure

K_E) is inversely related to the similarity coefficient of structural geometric parameters (here, the moment of inertia of the cross-section K_I) is inversely related, i.e., a softer model material implies a larger model structure size. SL plastics have a lower modulus of elasticity than aluminum alloys, so models made from them have a larger cross-sectional wall thickness.

The low modulus of the material is necessary to ensure the structural similarity of the dynamically similar model but also requires the support of processing technology printing capability. Bing-Shan Liu et al. designed and fabricated a box-section model of the wing based on low-modulus composite materials. The modal experiments show that the error range between the actual and target values of the first 2 orders of special diagnostic frequencies of this box-segment model is 4% ~ 15%, while the error of this book is less than 2%. This is because the SL process has a strong capability of printing internal and external structures, and is able to maintain a high machining accuracy for parts with irregular shapes and complex force transfer structures such as aircraft models, which is difficult to achieve with conventional processes.

In conclusion, the low modulus of elasticity of the materials used in SL and its powerful printing capability ensure the structural similarity of the dynamically similar model.

2) Weight reduction

From Table 9.5 it can be obtained that the total weight of the counterweight of the aluminum model is 15.4 g, which is much larger than the 1.3 g of its main structure, and the former is about 11.9 times of the latter, which makes the design and installation of the counterweight very difficult. On the contrary, the counterweight of the SL resin model is much smaller than its its main structure weight, which is only 0.1 times of the latter and almost negligible. Theoretically, zero counterweight can be achieved by fine-tuning the dimensional similarity factor (1:5 neighborhood).

9.4.2 Technical Limitations

1) Subjects of the case study

The object of this chapter is a box section without aerodynamic shape, and the correctness of the model implementation method is verified by means of modal testing. The box section is the main load-bearing member of the main wing of the aircraft, which plays a decisive role in the structural performance of the wing and the dynamic characteristics of the aircraft, so it is a representative object for the study; the modal test can be used to test the structural dynamic characteristics of the model, which is an inevitable step of the dynamic wind tunnel test.. However, the following research work is still needed to use the full SL resin dynamic model for the real aircraft wind tunnel test.

Strength design of all-SL resin model: A research team has designed a non-metallic dynamic model and conducted modal tests to verify that non-metallic materials can be used in the design of aircraft dynamic models.. A low modulus composite material was used in this study, which is similar to the SL resin used in this chapter in terms of elastic modulus and mechanical strength. However, no validation of wind tunnel tests was seen in this study, so the feasibility of non-metallic in wind tunnel tests still needs further study. One of the central questions is whether the strength of the non-metallic material is sufficient. Other research work in this book (see Chap. 5) shows that the all-resin F4 standard model is still safe at *Ma0.8 in* the FL-21 transonic wind tunnel (test section 0.6 m × 0.6 m), which provides a good basis for the subsequent research in this chapter.

Processing study of full SL resin model: The dynamic model is sensitive to the structural parameters of the model, so SL equipment is required to be able to process the complete model components at one time, such as processing a single wing at one time. Domestic commonly used transonic wind tunnel FL-24 (test Sect. 1.2m × 1.2 m) and FL-26 (test section 2.4 m × 2.4 m), the maximum size of the model used in the former is about 0.6 m ~ 0.8 m, and the latter is about 1.0 m ~ 1.5 m. The current processing size of mainstream SL printing equipment is in the order of 1 m [15] At present, the processing size of mainstream SL printing equipment is in the order of 1m, such as 3D system's iPro 9000XL SL equipment with a maximum print envelope of 1.5 m × 0.75 m × 0.55 m, and Xi'an Jiaotong University's SPS600B SL equipment with a maximum processing envelope of 0.6 m × 0.6 m × 0.45 m. The existing processing capacity can barely meet the requirements, but the country still needs to step up the development of large-capacity SL printing equipment in order to avoid being restricted in the key technology.

2) Efficiency issues of model design methods

The model design in this chapter is realized by the optimization calculation method. Because of the simple structure of the wing box itself and the simplification of the target vibration type selection (only the first two orders are taken), the optimization algorithm that comes with ANSYS can obtain satisfactory convergence results after a relatively short calculation time (about 100 calculation steps), which is sufficient for the research object of this chapter. However, the design problem of a real full-engine fluttering model is more complex and involves more counting optimization variables, and the optimization algorithm embedded in ANSYS appears to be less efficient. Therefore, special optimization algorithms need to be developed in order to improve the efficiency and accuracy of optimal design. Optimization design algorithms in the scope of structural dimensions at home and abroad have been developed to a more mature stage, and some researchers have conducted special algorithms for the design of flutter models, which can be referred to.

The torsional mode is an important inherent property of the wing, and the first two orders of the box section studied in this chapter do not involve torsion, so the torsional stiffness is not considered in the model stiffness design. In addition, the mass-similarity design in this chapter is based on the assumption that modal similarity and structural similarity automatically achieve modal similarity based on the joint determination of vibration similarity and solid frequency similarity. In a strict sense, this assumption needs to be tested. Therefore, the subsequent work should add the function of vibration similarity design and use the deformation extraction function of ANSYS to automatically obtain the nodal line information and optimize the calculation for its location configuration.

Chapter 10
Additive Manufacturing of Wind Tunnel Test Models with Pre-deformation

10.1 Overview

The wings of large transport aircraft generally have a large span to chord ratio, and the wing tip deformation under cruise conditions can reach the meter level. The huge difference in aerodynamic shape of the wing in different flight conditions makes it more difficult to accurately predict the aerodynamic performance of the aircraft and design accordingly, making this technology one of the key technologies for designing highly reliable and efficient large transport aircraft. In the conceptual design stage, a lot of calculations and experimental work are required in this regard. The computational techniques represented by multidisciplinary optimal design and fluid–structure coupling simulation technologies can help to solve many aerodynamic and structural mechanics problems in aircraft design [55] However, for higher-order aerodynamics and structural mechanics problems at high speeds, they can be used to solve the aerodynamic problems. However, for higher-order nonlinear aerodynamic problems (e.g., large deformations) at high speeds, wind tunnel testing techniques are widely considered to have higher reliability [58]. A series of rigid models of the aircraft and a static-elastic model are usually used in conducting wind tunnel tests. To improve the reliability of the data, the test data are first properly corrected, as shown in Fig. 10.1. The purpose of the static-elastic correction is to eliminate the data bias caused by the inevitable deformation of the rigid model during the blowing process. Of course, the correction process builds on the theoretical basis and a large amount of empirical data, which is time-consuming and labor-intensive, but indispensable under the current wind tunnel test program. Improvements in the design and processing techniques of wind tunnel test models can provide new implementations of existing wind tunnel tests that have the potential to improve this situation.

Because of its high accuracy and reliability, machining technology (represented by numerical control, NC) is the basis for the design and manufacturing of wind tunnel test models at present. However, recent advances in material science and processing technology have opened up the possibility of new model processing technologies,

W. Zhu and D. Li, *Models for Wind Tunnel Tests Based on Additive Manufacturing Technology*, Additive Manufacturing Technology,
https://doi.org/10.1007/978-981-99-5877-1_10

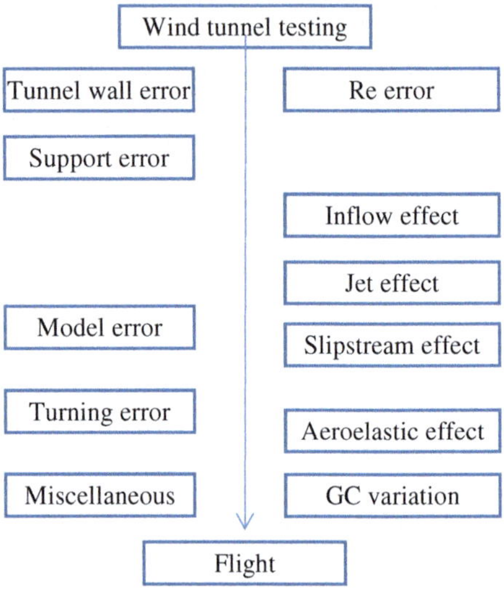

Fig. 10.1 Correction flow of wind tunnel test data

among which 3D printing (Additive Manufacturing, 3D printing) technology is the representative [7, 76]. The most recent advances in materials science and processing technology have opened up the possibility of new model processing technologies, including 3D printing (Additive Manufacturing). This technology was introduced into the manufacturing of wind tunnel test models at the early stage of development because of its powerful ability to process complex structures, and its feasibility has been fully demonstrated [12, 32, 33]. The feasibility of this technology has been well proven. However, the materials used in high-precision 3D printing processes are mostly non-metallic, represented by epoxy plastics. Because of the low strength and low modulus of these materials, the effectiveness of this technology in the manufacture of high-speed wind tunnel test models has been widely doubted, and for this reason various methods have been developed to strengthen the models made by 3D printing [21, 30, 55]. These researches are based on the idea of These studies are based on the idea that the weak mechanical properties of non-metallic materials used for 3D printing are a deficiency that needs to be avoided and overcome in the design and fabrication of wind tunnel test models. However, if a different viewpoint is taken and the characteristics of non-metallic materials are fully utilized, this "deficiency" may be transformed into an "advantage" and a completely new technical solution for wind tunnel testing may be developed as a result. In terms of mechanical strength, it is theoretically feasible to use non-metallic materials for the wind tunnel test model, considering that the stress on the model decreases dramatically after the size reduction, and the lower modulus of elasticity of non-metallic materials gives the model a higher deformation capacity, which allows it to deform as much as a real aircraft with a large span ratio. In addition, it is shown that the lower modulus of elasticity allows the model to retain a force transfer structure similar to that of a real aircraft. This is not possible for models machined from metal. At the same time, plastic materials have high damping, which is important for improving the safety of wind tunnel tests and the efficiency of data collection. The high damping characteristics of the material can reduce the model vibration during the blowing process, which helps to improve safety; meanwhile, the high damping allows the model to return to the stable position faster under the initial vibration, which reduces the acquisition waiting time and thus improves the data acquisition efficiency. In addition, the development of the integration of computational and experimental fluid dynamics also provides a basis for reassessing and improving existing aircraft model design and manufacturing techniques [1].

This book proposes a new technology for the design and fabrication of an aircraft wind tunnel test model with a large span-to-string ratio based on light-curing printing technology, providing a new wind tunnel test solution for the conceptual design of large aircraft. In this chapter, a flexible model with "initial negative deformation" is designed, which is deformed to the desired aerodynamic shape of the test during the blowing process under the load—the target shape—compared to the rigid model (metal). Compared with the rigid model (metal), this model can eliminate the correction of the elasticity effect of the metal model, and is closer to the actual aircraft in flight, achieving "deformation similarity". This chapter details the principles of this technique, addresses the key technologies involved, and uses a case study to demonstrate the feasibility of the approach.

10.2 Pre-deformation Design

10.2.1 Flexible Model

A large number of wind tunnel tests use rigid models. These models are generally made of high elastic modulus materials (e.g., steel, aluminum, etc.), but during the blowing process, the model inevitably deforms, as shown in Fig. 10.2.

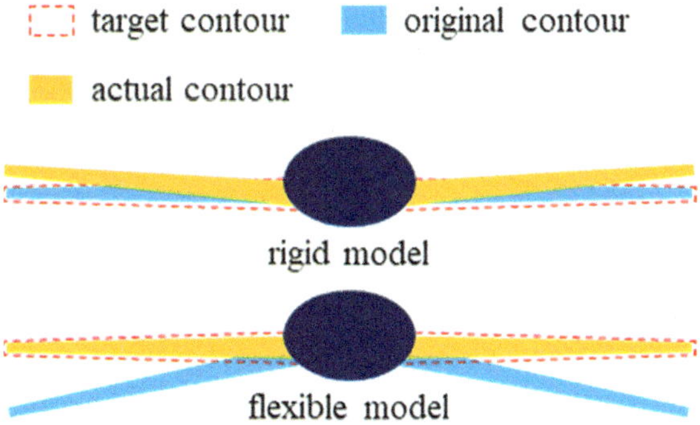

Fig. 10.2 Deformation of the rigid and elastic models in the test [43]

Since the model is assumed not to be deformed, the original contour of the model is the target contour, but under aerodynamic loading, the initial contour is deformed to the actual contour, thus deviating from the target contour. In order to eliminate the influence of model deformation on data accuracy, the collected data are generally subjected to correction of deformation.

In order to avoid the extra cost and time consuming and possible risk of accuracy loss caused by deformation correction, a flexible model with "initial negative deformation" can be designed, and the blowing process will deform to the desired aerodynamic shape under the action of the load—the target shape, eliminating the deformation correction and directly obtaining the test data under the shape. In the process of blowing, the air is deformed to the desired aerodynamic shape under the load—the target shape, eliminating the deformation correction, and directly obtain the test data under the shape. As shown in Fig. 10.2, the model with "negative deformation" of the initial shape, the actual shape after deformation under the aerodynamic load and the target shape coincide.

Machining techniques to support complex stiffness distribution models and calculation methods to obtain initial negative deformations are the key technologies of this new method. The rigid model requires sufficient stiffness and qualitative accuracy, while the flexible model requires quantitative stiffness distribution to a specific design value (similar to the aeroelastic model), which requires higher design methods and machining techniques. If the traditional rigid model uses the CNC machine tool-based processing technology, the cost and cycle time will be greatly increased, offsetting the benefits of the flexible model from deformation correction, which is not economically feasible. At the same time, the traditional model processing material is mainly metal, its elastic modulus is large, compared with low modulus material, its deformation range is smaller, the same pneumatic load, the pre-deformation is smaller. The development of 3D printing technology represented by SL process provides another option. As mentioned above, the SL process technology is based on a thermosetting plastic and is capable of economical machining of parts with complex internal and external structures with high precision. A comparison of the performance parameters of SL plastics and two commonly used model materials is shown in Table 1.3. On the one hand, the fast processing capability of complex internal and external structures provides the possibility of realizing complex stiffness distribution of flexible models and reduces the difficulty of pre-deformation calculation and model design; on the other hand, low-modulus plastics have a large deformation capability, which provides a larger design range for pre-deformation calculation. At the same time, compared with metallic materials, plastics have larger damping, which is beneficial to attenuate the model vibration caused by wind tunnel blowing fluctuations.

10.2.2 Pre-deformation Calculation

As mentioned above, the acquisition of the pre-deformation amount is the key to the design of flexible model. In this book, the model deformation under specific aerodynamic load is accurately obtained by using the fluid–solid coupling calculation method, and based on this, the amount of deformation with the flexible model under certain working conditions is obtained with the help of optimization method.

1) Acquisition of model aerodynamic deformation—fluid–solid coupling calculation

CFD/CSD coupling method is the core technology of deformation calculation in this technology, which is usually used for high precision aeroelastic analysis [21]. This book uses CFD and CSD to calculate the mathematical model respectively as discussed below.

The CFD solved fluid control equation as Eq:

$$\frac{\partial}{\partial t}\iiint\limits_{V} \boldsymbol{Q} dV + \iint\limits_{\partial V} \boldsymbol{F}\cdot\boldsymbol{n} dS = 0 \tag{10.1}$$

where $\boldsymbol{Q} = [\rho, \rho u, \rho w, \rho e]^T$, ρ, u, v, w, e *are the* air density, the components of the velocity vector in the three coordinate directions and the total internal energy per unit mass, respectively. $\boldsymbol{n}$ is the normal unit vector of the area fraction, and V is the volume fraction domain, and ∂V is the boundary containing the volume fraction, and $\boldsymbol{F}$ is the flux term, including the inviscid term $\boldsymbol{F}_E$ and the viscous term $\boldsymbol{F}_\nu$. If the viscous term is considered $\boldsymbol{F}_\nu$, Eq. (10.1) is the Navier-Stoks equation, and if the viscous term is not considered $\boldsymbol{F}_\nu$, Eq. (10.1) is the Euler equation. Flux term $\boldsymbol{F}$ is discretized using the central format and the time discretization is done by the dual time method [32].

The structural motion control equation solved by CSD is as the Eq. (6.13). Without considering the effect of velocity and acceleration, the Eq. (6.13) can be simplified to the following equation, which is the structural static equilibrium equation.

$$[K]\{q(t)\} = F(t) \tag{10.2}$$

According to the way of coupling, CFD/CSD coupling can be divided into three main categories: full coupling, loose coupling and tight coupling [77]. In the fully coupled model approach, the fluid equations and the structural equations of motion are coupled in the same way. In the fully coupled model approach, the fluid equations and structural equations of motion are represented in a unified form, and the fluid equations and structural equations of motion are solved simultaneously in physical time. The fully coupled method is computationally expensive, limited by the grid size, and is only used to solve 2D simple flow problems. In the loosely coupled model approach, the fluid and structural equations of motion are solved by separate solvers, and external data exchange is used between the flow and structural modules, i.e., data exchange starts only after partial or full convergence. This type of data exchange increases the flexibility in the selection of the solution code, but the computational accuracy is low and limited to solving problems with small disturbances and moderate nonlinearities. In the Closely coupled model approach, the flow field and structural equations of motion are solved by different solvers, but are coupled in a single module, and data are exchanged at the interface or boundary through an interface technique. The tightly coupled method uses only internal data exchange between the flow field module and the structure module, i.e., the data exchange starts when the data are not converged, and can be used to solve problems with large perturbations and high nonlinearity. The data exchanged here refers to the aerodynamic loads mapped from the fluid surface mesh to the structural dynamics mesh and the structural displacements mapped from the structural dynamics mesh to the CFD surface mesh.

The book adopts a tightly coupled model, the flow field is discretized to solve the non-constant N-S equation, the structural field is discretized to solve the structural static equilibrium equation, the dynamic mesh is updated by the spring-like extrapolation method, and the interface data is transferred by the Virtual Surface (VS) method.

2) Obtain model pre-deformation—Optimization calculation

The implementation of the model design is shown in Fig. 10.3.The purpose of the design is to obtain the "design shape" of the model, which has an additional shape and can be deformed to the "target shape" under test conditions (defined as the "design point" of the model) under load. The "target shape" is the shape of the aircraft. The "target profile" is an aerodynamic state of the aircraft, usually the aerodynamic optimum in the aircraft design, which is expected to have better aerodynamic performance at the "design point", while its real aerodynamic characteristics need to be verified by wind tunnel tests.

If the model is realized using metal machining methods, then the shape is the "machined shape" of the model. Metal models are considered rigid and ideally require "zero deformation" during the blowing process, however, this is not possible in practice. In order to fix the errors caused by this unavoidable deformation, the test data needs to be "deformation corrected". In contrast to metal machined models, this book uses 3D printing technology to produce non-metal (epoxy plastic) "flexible models". This type of model has the same large deformation capacity as a real aircraft in the blowing state. If the shape of the model has a reverse "pre-deformation", using its large deformation capacity, it can be deformed to the "target shape" in the blowing state, and the required experimental data can be obtained directly without correction.

The core of this new technique is how to obtain the "design shape" from the "target shape". This book is based on the coupled method of Computational Fluid Dynamics and Computational Structural Dynamics (CSD), using optimization calculations to obtain the shape. The optimization mathematical model is given by the following equation:

$$\text{minimize} f(\boldsymbol{x}) \tag{10.3}$$

where the objective function $f(\boldsymbol{x})$ is defined as the deviation of the "acquired profile" from the "target profile". Theoretically, the shape deviation should be calculated by considering all deviations of all nodes of all shapes, but in engineering practice, this method is too large to be feasible. In this book, we adopt a simplified strategy by

Fig. 10.3 Schematic diagram of calculation of pre-deformation amount

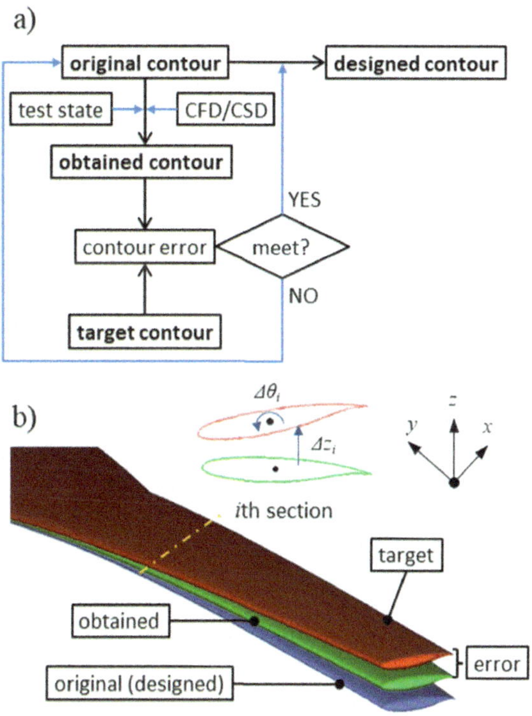

(a) Calculation flow (b) Shape relationship

selecting only the critical phases of the shape and the deviations that play a key role in the aerodynamic characteristics. As shown in Fig. 10.3 (b), several critical sections of the wing (such as the *ith* section in the figure) are selected and only their translational deviations along the z-direction and rotational deviations around the torsional center are calculated in this book. The simplified objective function can be expressed as the following equation, where the α_θ and α_z are the corresponding computational weights.

$$f(\boldsymbol{x}) = \alpha_\theta \sum_i^n |\Delta\theta_i| + \alpha_z \sum_i^n |\Delta z_i| \tag{10.4}$$

The vectors in the objective function $\boldsymbol{x}$ are optimization variables that define a specific set of "initial profiles". When the objective function $f(\boldsymbol{x})$ is smaller than the set threshold, the "initial shape" is the "design shape" of the optimization calculation, and the "machining shape" of the model is obtained, i.e., after convergence of the optimization The "initial shape" = "design shape" = "processing shape". As mentioned above, the "initial shape" is obtained from the "target shape" by "inverse pre-deformation", and the optimization variables $\boldsymbol{x}$ defines the amount of deformation. As with the simplified method of obtaining the objective function described above, the optimization variables $\boldsymbol{x}$ are also composed in a similar way:

$$\boldsymbol{x} = (z_1, z_2, \ldots, z_n, \theta_1, \theta_2, \ldots, \theta_{\mathrm{n}}) \tag{10.5}$$

Based on the coupled CFD/CSD calculation method, the static elastic deformation of the model of a specific configuration under aerodynamic load can be predicted. Based on this method, the optimization idea is that the initial shape of the model is used as the design target, and the modified shape of the model is found so that the shape after deformation under aerodynamic load of specific conditions is the same as the initial shape, so that the aerodynamic data obtained removes the influence of elastic deformation. At this time, the deformation correction shape is the manufacturing shape of the model.

Figure 10.3a gives the block diagram of the optimization calculation, which can be used to find the manufacturing profile of the model by the method of finding the optimization, where the error between the deformed profile and the initial profile is judged both quantitatively and in terms of the size of the aerodynamic impact of the profile error. The error source analysis sets the corresponding basic criteria to ensure the rapidity of the optimization search, such as the change in profile angle of orientation generated by torsional deformation has a greater effect on lift, drag and pitch moment, while the bending deformation has a smaller effect on force, etc.

10.3 Case Studies

10.3.1 Model Design

The F4 model is a standard model of a wing-fuselage assembly with a large span-to-skin ratio supercritical wing, which was used to study the relevance of experimental data from three large European transonic wind tunnel civil aircraft and to verify the usability of the ternary transonic wing-fuselage assembly calculation program [78]. The F4 model has a fuselage length of $L = 1192$ mm and a wingspan of $b = 1171.29$ mm. The cross-sectional dimensions of the wind tunnel in which the validation experiments are conducted are 0.6 m × 0.6 m, so the F4 model is designed and manufactured at a scale of 1:4. SL resin material was used (for reference of performance parameters). The five sets of F4 manufacturing profiles were designed for machining under the design conditions (Mach number $Ma = 0.6$, angle of approach $\alpha = 0°$, velocity pressure $q = 20$ kPa) according to the deformation correction method. The five sets of F4 manufacturing profiles are shown in Fig. 10.4, where No. 3 is the "design profile" and No. 1, 2, 4 and 5, which are symmetrically distributed on the top and bottom, are designed for comparison purposes.

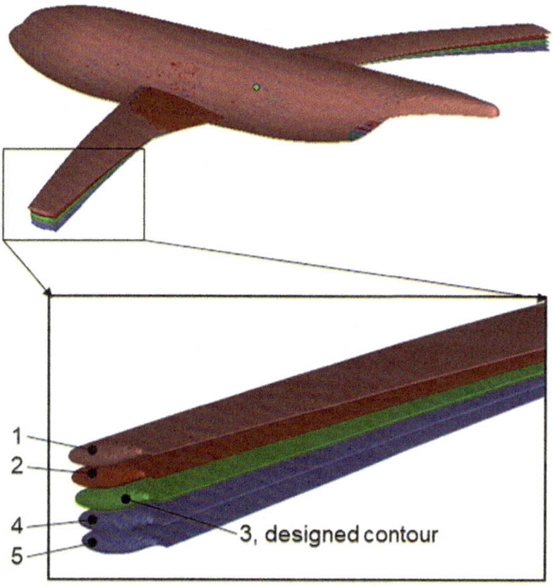

Fig. 10.4 Model manufacturing shape and initial shape

The aerodynamic parameters for the strength check are as follows Table 10.1 where the angle of attack is chosen as the maximum value of 2°.

Stress distribution of F4 resin model under aerodynamic action at design conditions Fig. 10.5 The distribution is as expected.

As shown in the figure, the maximum stress value occurs in the middle position of the wing, which is 15 MPa, less than the material allowable stress of 23 MPa, indicating that the designed light-cured F4 model meets the strength requirements of the wind tunnel test.

In order to measure the aerodynamic forces acting on the model, the model needs to be mounted on a measurement mechanism in the wind tunnel, and the model is connected to the balance by means of a transition taper sleeve and pins, as shown in Fig. 10.6.

Table 10.1 Aerodynamic parameters and wind tunnel test conditions for strength calibration

Parameters	Value
Mach number *Ma*	0.6
Angle of attack α/ (°)	-2, ~2
Yaw angle β/ (°)	0
Rolling angle γ/ (°)	0
Reynolds Number *Re*	$4.3 \times 10e^5$
Total pressure $P0_{\infty}$ /kPa	101

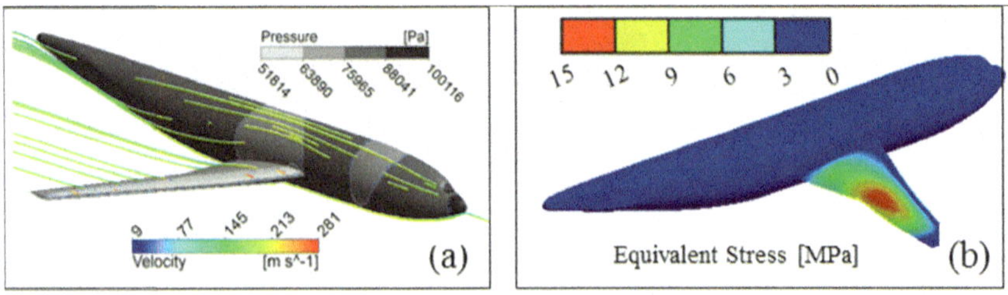

(a) Results of flow field analysis (b) Stress distribution

Fig. 10.5 Strength calibration results

Fig. 10.6 Model structure diagram

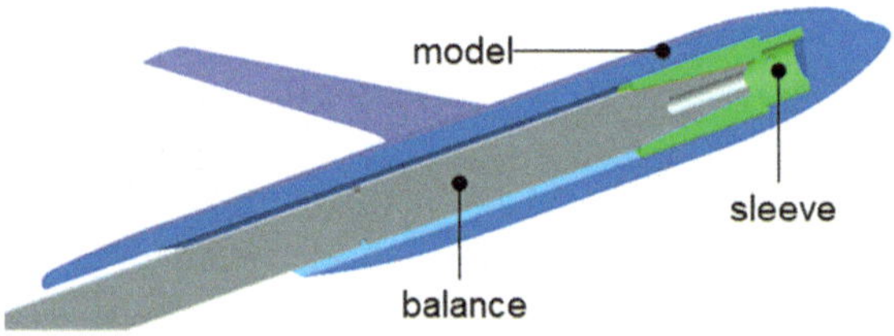

10.3.2 Forming and Manufacturing

After encryption and smoothing, 5 sets of F4 models were processed on the SPS600B light-curing rapid prototyping machine.

In order to reduce the influence of the step effect of each section of light-curing rapid forming on the surface of the wing on the surface of the model, the forming direction is chosen in the direction of the model axis, and the support is loaded on the head and leading edge of the wing. However, because the head and wing leading edge supports are point supports, and the support height is too high, the support stiffness is low, and the support deformation is easily generated during processing, which reduces the forming accuracy and even leads to the scrapping of the processed model, so the two models are processed simultaneously, and the support is added between the back of the fuselage of the model to increase the support stiffness, as shown in Fig. 10.7.

The surface roughness of the model is 1.4 μm after the support removal and grinding process, which meets the surface roughness requirement of the model for wind tunnel testing. Although the forming accuracy of the light-curing rapid forming method is relatively low compared with the machining method, compared with the traditional machining manufacturing method, the light-curing rapid forming method can reduce the processing cycle and processing cost, and it is easy to realize the seamless one-piece processing of small models without model assembly, and the overall performance of the processed model is good.

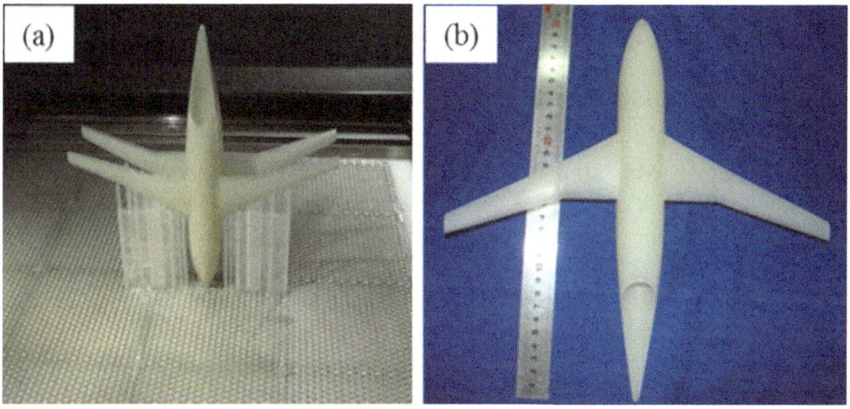

Fig. 10.7 Model forming direction and support position

10.3.3 Wind Tunnel Test

The model is installed in the wind tunnel as shown in Fig. 10.8.

The wind tunnel test of the model was conducted in the FL-21 trans-hypersonic wind tunnel of the High Speed Institute of the China Aerodynamic Research and Development Center, and the dimensions of the FL-21 wind tunnel test section were 0.6 m × 0.6 m × 1.75 m. The balance was a six-component balance with the number 2N6-18A. The wind tunnel test conditions were as follows Table 10.1, where Reynolds number Re is the Reynolds number based on the average aerodynamic chord length of the wing.

Fig. 10.8 Installation diagram of the model in the wind tunnel

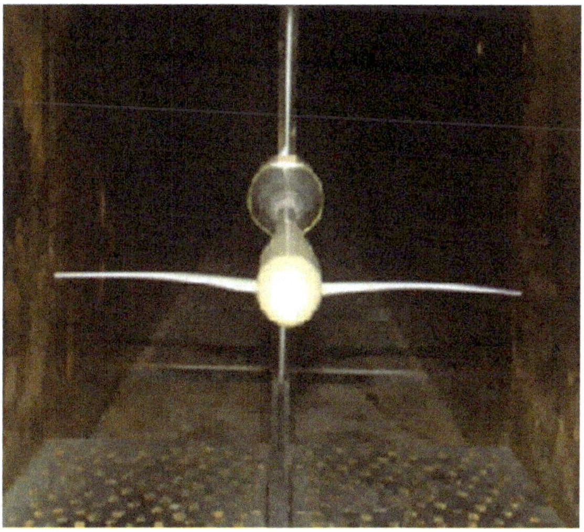

10.4 Analysis and Discussion

10.4.1 Technology Availability

For Mach number Ma of 0.6, the lift coefficient C_L and drag coefficient C_D relative to the angle of attack α curve are shown in Fig. 10..9. Qualitatively, all C_L -α curves are approximately linearly distributed with the same trend as that obtained from the metal model [14].

Table 10.2 shows the lift coefficients C_L and drag coefficients C_D for the different models at the "design point". The data for the metal models are from three European wind tunnels—NLR, ONERA and DRA—and are corrected for hydro-elasticity and deformation [78]. The resin model used in this test was obtained in the FL-21 wind tunnel without hydro-elastic and deformation corrections. As expected, the lift coefficient CL = 0.445 for the model with the "design profile" (#3) was the closest to the average value of 0.440 in the three wind tunnels. Among all the results of this test, model #3 with the design profile has the largest lift coefficient C_L, as expected, because of the optimal aerodynamic characteristics of the modified profile in the design phase [14]. At the same time, the lift coefficients of models 1 and 2 are closer, and the data of models 4 and 5 have similar characteristics, which is reasonable due to the fact that in the model design is the shape of the above two groups of models are located on the upper and lower surfaces of No. 3 (as Fig. 10.4 shown).

For the drag characteristics, the results of this test differed significantly from those of the three wind tunnels are shown in Fig. 10.9 and Table 10.2. First, the drag coefficient C_D is one order of magnitude smaller than the lift coefficient C_L, so the drag coefficient C_D was not considered in the model shape optimization design. Another reason is the difference of Reynolds number in the wind tunnel test. The Reynolds number in this test is $Re = 4.3 \times 10^5$, and $Re = 3 \times 10$ in the literature[6] [78], with an order of magnitude difference between the two. Because the Reynolds number *Re is a* dimensionless coefficient characterizing the viscous influence of fluid medium, its smaller value means the greater the viscous influence and thus the greater the drag. This is another reason why the drag coefficients C_D in this test are greater than those reported in the literature.

Table 10.2 Comparison of basic aerodynamic coefficients

Testing	Resin model, FL-21 tunnel					Metal models, other wind tunnels			
	1	2	3	4	5	NLR	ONERA	DRA	Average
C_L	0.404	0.391	0.445	0.428	0.430	0.443	0.441	0.437	0.440
C_D	0.0314	0.0305	0.0319	0.0321	0.0314	0.0258	0.0251	0.0249	0.0253

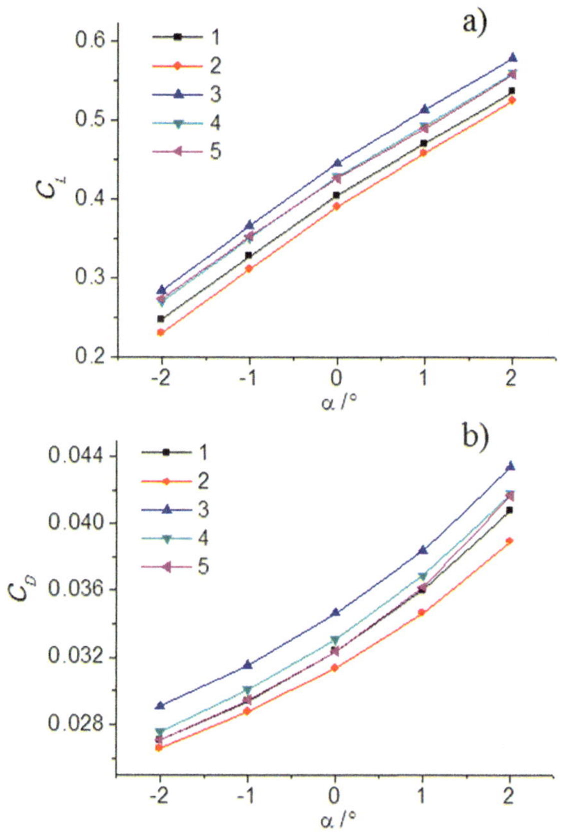

Fig. 10.9 Variation of model aerodynamic coefficient with angle of attack α (Ma = 0.6)

10.4.2 Technical Limitations

1) Subjects of the case study

The F4 model is a standard model of a wing-body assembly with a fuselage length of 1.192 m, a wingspan of 1.171 m, and a large span-to-skin ratio supercritical wing, and was used to study the correlation of experimental data from three large European transonic wind tunnel civil aircraft and to verify the usability of the ternary transonic wing-body assembly calculation program; the wind tunnel test of the model was conducted in the FL-21 transonic wind tunnel at the China Aerodynamic Research and Development Center. The dimensions of its test section are 0.6 m × 0.6 m × 1.75 m; the scale of the model is 1:4, with 298 mm length and 292 mm wingspan, and the full model can be processed at one time using SPS600B SL printing equipment. The case study made in this chapter is a typical representative of a large span-to-string ratio aircraft prototype, a simplified structural model, and a small wind tunnel test model. In engineering applications, a large number of force tests are conducted in larger wind tunnels (e.g., FL-26 wind tunnel with a test section of 2.4 m × 2.4 m), the model size is large (1 m magnitude), and the model is rich in detail features (e.g., including adjustable rudder surfaces, refer to Chap. 2), which is not easy to be processed at one time using SL equipment, which will certainly increase the difficulty of pre-deformation design and model processing, and requires targeted research. This method is suitable for the selection design in the conceptual design stage of large transport aircraft, and can be used to design and manufacture a large number of wind tunnel test models with a range of shapes, and conduct a large number of wind tunnel tests to select the solution with the best aerodynamic shape.

2) General pre-deformation design method

The F4 model in this chapter only considers the wing deformation, and the wing is a solid structure, so the design of the model does not need to consider the deformation of the fuselage or the tail section and the design of the internal structure of the model. However, for larger size and force measurement models with detailed features, it is necessary to consider the deformation of parts other than the wing in the pre-deformation design, and to add the structural design function in the model design. For the deformation of other parts, we can refer to the optimization calculation method of wing deformation in this chapter, but we need to consider the methods to improve the efficiency of CFD/CSD coupled deformation calculation and pre-deformation optimization algorithm; the design of internal structure of the model can refer to the design methods in other chapters of this book (such as the optimization design of model stiffness in Chaps. 3 and 4).

References

1. Marcus, S., L. Tommie, and P. William. 2004. Integrated test and evaluation for the 21st century. Woodland Hills, California.
2. Fujii, K. 2005. Progress and future prospects of CFD in aerospace—Wind tunnel and beyond. *Progress in Aerospace Sciences* 41 (6): 455–470.
3. Zhu, W. 2019. Models for wind tunnel tests based on additive manufacturing technology. *Progress in Aerospace Sciences.*
4. Gionet, C., and F. Yandrasts. 1997. Developments in model design and manufacturing techniques. Royal Aeronautical Society 1997 European Forum, Wind Tunnels and Wind Tunnel Test Techniques.
5. Springer, A., K. Cooper, and F. Roberts. 1997. *Application of rapid prototyping models to transonic wind-tunnel testing*. Reno, NV: AIAA 97-0988.
6. Timothy, G. 2008. Process improvement for aerodynamic model development. In *AIAA Modeling and Simulation Technologies Conference and Exhibit*. Honolulu, Hawaii: AIAA.
7. Heyes, A.L., and D.A.R. Smith. 2004. Rapid technique for wind-tunnel model manufacture. *Journal of Aircraft* 41 (2): 413–415.
8. Black, C., K.V. Singh, S. Goodman, et al. 2018. *Design, fabrication and testing of 3D printed wings for rapid evaluation of aeroelastic performance*, 1–17. Kissimmee, Florida: AIAA.
9. Springer, A., K. Cooper, and F. Roberts. 1997. *Application of rapid prototyping models to transonic wind-tunnel testing*. Reno, NV: AIAA.
10. Springer, A., and K. Cooper. 1997. *Comparing the aerodynamic characteristics of wind tunnel models produced by rapid prototyping and conventional methods.* Atlanta, GA, US: AIAA.
11. Chuk, R.N., and V.J. Thomson. 1998. A comparison of rapid prototyping techniques used for wind tunnel model fabrication. *Rapid Prototyping Journal* 4 (4): 185–196.
12. Springer, A. 1998. Evaluating aerodynamic characteristics of wind-tunnel models produced by rapid prototyping methods. *Journal of Spacecraft and Rockets* 35 (6): 755–759.
13. Bird, J. 2012. Exploring the 3D printing opportunity. *The Financial Times London.*
14. House, T.W. 2012. *We can't wait: Obama administration announces new public-private partnership to support.* Washington, DC: The White House Office of the Press Secretary.
15. Wohlers, T., and T. Caffrey. 2012. *Wohlers report 2012.* Ft. Collins: Wohlers Associates, Inc.
16. Springer, A., and K. Cooper. 1997. *Comparing the aerodynamic characteristics of wind tunnel models produced by rapid prototyping and conventional methods.* AIAA.
17. Reeder, M.F., W. Allen, J.M. Phillips, et al. 2008. Wind-tunnel measurements of the E-8C modeled with and without winglets. *Journal of Aircraft* 45 (1): 345–348.
18. Butler, T.W. 2005. *Model design and fabrication improvements for wind tunnel testing*, 363–369. Nashville, TN, United States: American Institute of Aeronautics and Astronautics Inc.
19. Bartley-Cho, Jonathan, and J. Henderson. 2008. Design and analysis of HiLDA/AEI aeroelastic wind tunnel model. Honolulu, Hawaii.

W. Zhu and D. Li, *Models for Wind Tunnel Tests Based on Additive Manufacturing Technology*, Additive Manufacturing Technology,
https://doi.org/10.1007/978-981-99-5877-1

20. Romeo, G., G. Frulla, Enrico Cestino, Pier Marzocca, and Ilhan Tuzcu. 2006. Nonlinear aeroelastic modeling and experiments of flexible wings. Newport, Rhode Island.
21. Tyler, C., W. Braisted, and J. Higgins. 2005. *Evaluation of rapid prototyping technologies for use in wind tunnel model fabrication*, 5081–5088. Reno, NV, United States: AIAA.
22. Tyler, C. 2004. *A joint computational fluid dynamics and experimental fluid dynamics test program*, 11270–11279. Reno, NV, United states: American Institute of Aeronautics and Astronautics Inc.
23. Buck, G.M. 2000. Rapid model fabrication and testing for aerospace vehicles. Reno, Nevada.
24. Tyler, C., M. Reeder, W. Braisted, et al. 2004. Rapid technology focused experimental and computational aerodynamic investigation of a strike tanker.
25. Hildebrand, R.J., R. Eidson, and C. Tyler. 2003. Development of a low cost, rapid prototype, lambda wing-body wind-tunnel model. Orlando, Florida.
26. Landrum, D.B., Robert M. Beard, Paul A. LaSarge, and Natalie von Sprecken. 1997. *Evaluation of stereolithography rapid prototyping for low speed airfoil design*. Reno, NV: AIAA.
27. Azarov, Y.A., V.D. Vermel, A.V. Kornushenko, et al. 2002. *Experience in laser stereolithography and its application in manufacturing wind-tunnel aerodynamic models of various purposes*, 433–440. Suzdal, Russia: SPIE.
28. Nadooshan, A.A., S. Daneshmand, and C. Aghanajafi. 2007. *Application of RP technology with polycarbonate material for wind tunnel model fabrication*, 1–6. Bangkok, Thailand: World Academy of Sciences, Engineering & Technology-Waset.
29. Zhou, Z., D. Li, J. Zeng, et al. 2007. Rapid fabrication of metal-coated composite stereolithography parts. *Proceedings of the Institution of Mechanical Engineers, Part B: Journal of Engineering Manufacture* 221 (9): 1431–1440.
30. Zhou, Z., D. Li, Z. Zhang, et al. 2008. Design and fabrication of a hybrid surface-pressure airfoil model based on rapid prototyping. *Rapid Prototyping Journal* 14 (1): 57–66.
31. Dang-guo, Yang, Zhang Zheng-yu, Sun Yan, Zhu Wei-jun. 2011. A preliminary design and manufacturing study of hybrid lightweight high-speed wind-tunnel models. *Rapid Prototyping Journal* 17 (1): 10.
32. Landrum, D.B., R.M. Beard, P.A. LaSarge, et al. 1997. Evaluation of stereolithography rapid prototyping for low speed airfoil design. Reno, NV.
33. Rn, C., and T. Vj. 1998. A comparison of rapid prototyping techniques used for wind tunnel model fabrication. *Rapid Prototyping Journal* 4 (4): 185–196.
34. Hildebrand, R.J., R.C. Eidson, and C. Tyler. 2003. Development of a low cost, rapid prototype, lambda wing-body wind-tunnel model. Orlando, Florida.
35. Wohlers, T. 2008. *Wohlers report 2008 executive summary*. America: Wohlers Associates, Inc.
36. Fujino, M., H. Oyama, and H. Omotani. 2003. *Flutter characteristics of an over-the-wing engine mount business-jet configuration*, 4925–4936. Norfolk, VA, United States: AIAA.
37. Junk, S., W. Schröder, and S. Schrock. 2017. Design of additively manufactured wind tunnel models for use with UAVs. *Procedia CIRP* 60: 241–246.
38. Tyler, C., M.F. Reeder, W. Braisted, et al. 2004. Rapid technology focused experimental and computational aerodynamic investigation of a strike tanker. AIAA.
39. Jamieson, R. 1990. New options in aerodynamic modelling: CAD, CFD, stereolithography and shading in a workstation environment. *Computer-Aided Engineering Journal* 7 (3): 75–79.
40. Zhu, W.J., D.C. Li, Z.Y. Zhang, et al. 2011. Design and fabrication of stereolithography-based aeroelastic wing models. *Rapid Prototyping Journal* 17 (4): 298–307.
41. Tyler, C., Richard J. Schwartz, G. Fleming, et al. 2008. *Rapid prototyping-unmanned combat air vehicle (UCAV)/sensorcraft*. Air Force Research Laboratory.
42. Reeder, M.F., W. Allen, J.M. Phillips, et al. 2007. Wind-tunnel measurements of the E-8C modeled with and without winglets. *Journal of Aircraft* 45 (1): 345–348.
43. Zhu, W., X. Zhang, and D. Li. 2019. Flexible all-plastic aircraft models built by additive manufacturing for transonic wind tunnel tests. *Aerospace Science and Technology* 84: 237–244.
44. Kampchen, M., A. Dafnis, H.G. Reimerdes, et al. 2003. Dynamic aero-structural response of an elastic wing model. *Journal of Fluids and Structures* 18 (1): 63–77.

45. Wang, C., G. Yin, Z. Zhang, et al. 2015. Design and fabrication of an aircraft static aeroelastic model based on rapid prototyping. *Rapid Prototyping Journal* 21 (1): 34–42.
46. Korsch, H., A. Dafnis, and H.G. Reimerdes. 2009. Dynamic qualification of the HIRENASD elastic wing model. *Aerospace Science and Technology* 13 (2–3): 130–138.
47. ISO. 2019. Determination of tensile properties—Part 1: General principles. Plastics. ISO.
48. ISO. 2012. Determination of tensile properties—Part 2: Test conditions for moulding and extrusion plastics. Plastics. ISO.
49. ISO. Determination of flexural properties. Plastics. ISO.
50. Saleh, N. 2004. Effects of electroplating on the mechanical properties of stereolithography and laser sintered parts. *Rapid Prototyping Journal* 10 (5): 305–315.
51. Azarov, Y.A., V.D. Vermel, A.V. Kornushenko, et al. 2002. *Experience in laser stereolithography and its application in manufacturing wind-tunnel aerodynamic models of various purposes*, 433–440. Suzdal, Russia: SPIE.
52. Tyler, C., R.J. Schwartz, G. Fleming, et al. 2008. *Rapid prototyping-unmanned combat air vehicle (UCAV)/sensorcraft*. Air Force Research Laboratory.
53. Aghanajafi, C., S. Daneshmand, and A.A. Nadooshan. 2009. Influence of layer thickness on the design of rapid-prototyped models. *Journal of Aircraft* 46 (3): 981–987.
54. Heisler, R.R., and C.L. Ratliff. 2001. Wind tunnel model design and testing using rapid prototype materials and processes. In *The 10th Annual AIAA/BMDO Technology Conference*, Williamsburg, Virginia.
55. Bhardwaj, M.K., R.K. Kapania, E. Reichenbach, et al. 1998. Computational fluid dynamics/ computational structural dynamics interaction methodology for aircraft wings. *AIAA Journal* 36 (12): 2179–2186.
56. Marvin, J.G. 1994. Book review turbulence modeling for CFD. *AIAA Journal* 32 (6): 1339–1339.
57. Chen, P., and I. Jadic. 1998. Interfacing of fluid and structural models via innovative structural boundary element method. *AIAA Journal* 36 (2): 282–287.
58. Fujii, K. 2005. Progress and future prospects of CFD in aerospace—Wind tunnel and beyond. *Progress in Aerospace Sciences* 41 (6): 455–470.
59. Wendt, J.F. 2009. *Computational fluid dynamics*, chaps. 3, 4, 8. Berlin: Springer.
60. Selin, A., and K. Doyle. 2004. Simulation of supersonic cavity flow using 3D RANS equations. In *22nd Applied Aerodynamics Conference and Exhibit*. Providence, Rhode Island: AIAA.
61. Thorby, D. 2008. *Structural dynamics and vibration in practice: An engineering*, chaps. 6, 7, 8. Amsterdam Boston Elsevier/Butterworth-Heinemann.
62. Scanlan, Robert H., and Robert Rosenbaum. 1951. *Introduction to the study of aircraft vibration and flutter*. New York: Macmillan Company.
63. Jewel B. Barlow, W.H. Rae, Jr., and Alan Pope. 1999. *Low-speed wind tunnel testing*, chaps. 1, 11, 12. New York: Wiley.
64. Bettina, W., R. Dominik, K. Florian, et al. 2008. Additive processing of polymers. *Macromolecular Materials and Engineering* 293 (10): 799–809.
65. Stotko, C.M. 2009. Laser sintering: Layer by layer. *Nature Photonics* 3 (5): 265–266.
66. Zhu, W.J., K. Miao, and D.C. Li. 2019. Static aeroelastic models with integrated stiffness-contributing shell structures built by additive manufacturing. *Engineering Structures* 187: 352–361.
67. French, M., and F.E. Eastep. 1996. Aeroelastic model design using parameter identification. *Journal of Aircraft* 33 (1): 198–202.
68. Raja, S., S. Rajappa, and S. Janardhanam. 2009. Aeroelastic testing of aerospace vehicles—Experiences gained during four decades. In *National Conference on Wind Tunnel Testing (2009-IT11)*, IISc, Bangalore, India.
69. Chedrik, V.V., F.Z. Ishmuratov, M.C. Zichenkov, et al. 2004. *Optimization approach to design of aeroelastic dynamically-scaled models of aircraft*, 3812–3823. Albany, NY, United States: American Institute of Aeronautics and Astronautics Inc.
70. Zeng, D., Y. Yan, B. Liu, et al. 2006. Design of low-velocity wing tunnel flutter model of aircraft composite structure. *Acta Aeronautica et Astronautica Sinica* 27 (2): 232–235.

71. Li, Shaohua, Z. Yang, and Yingsong Gu. 2009. Partial structure similarity method to design composite transonic flutter model. *Journal of Mechanical Strength* 31 (2): 22339–22343.
72. Dornfeld, W.H. 1995. Direct dynamic testing of scaled stereolithographic models. *Sound and Vibration* 28–29 (8): 12–17.
73. Ziemian, C., R. Ziemian, and E. Barker. 2010. Shake-table simulation study of small scale layered models. *Rapid Prototyping Journal* 16 (1): 4–11.
74. Mahn, J.P., and P.V. Bayly. 1999. Impact testing of stereolithographic models to predict natural frequencies. *Journal of Sound and Vibration* 224 (3): 411–430.
75. Heeg, J., C.V. Spain, and J.A. Rivera. 2004. Wind tunnel to atmospheric mapping for static aeroelastic scaling. In *5th AIAA/ASME/ASCE/AHS/ASC Structures, Structural Dynamics and Materials Conference*. Palm Springs, California: AIAA.
76. Ciancarelli, C., M. Melanson, and W. Baker. Wind tunnel test technologies for aircraft cycle time reduction. In *37th AIAA Aerospace Sciences Meeting and Exhibit*. Reno, Nevada: AIAA.
77. Friedmann, P.P. 1999. Renaissance of aeroelasticity and its future. *Journal of Aircraft* 36 (1): 105–121.
78. Development AGfAR. 1994. *A selection of experimental test cases for the validation of CFD codes*. Neuilly-sur-Seine, France: NATO Research and Technology Organisation.

SPRINGER NATURE

GPSR Compliance

The European Union's (EU) General Product Safety Regulation (GPSR) is a set of rules that requires consumer products to be safe and our obligations to ensure this.

If you have any concerns about our products, you can contact us on ProductSafety@springernature.com

In case Publisher is established outside the EU, the EU authorized representative is:

Springer Nature Customer Service Center GmbH
Europaplatz 3
69115 Heidelberg, Germany

The manufacturer's authorised representative in the EU is Springer Nature Customer Service Centre GmbH, Europaplatz 3, 69115 Heidelberg, Germany. If you have any concerns regarding our products, please contact ProductSafety@springernature.com

Printed and bound by CPI Group (UK) Ltd, Croydon, CR0 4YY
01/10/2026
02225867-0001